# Numberama
# Recreational Number Theory
# In The School System

## Authored by

## Elliot Benjamin, Ph.D., Ph.D.

*Instructor of Mathematics at CAL Campus; Psychology Mentor,*
*Ph.D Committee Chair at Capella University, Minneapolis, USA*

**Numberama: Recreational Number Theory In The School System**

Author: Elliot Benjamin Ph.D. Ph.D

eISBN (Online): 978-1-68108-512-8

ISBN (Print): 978-1-68108-513-5

## General:

1. Any dispute or claim arising out of or in connection with this License Agreement or the Work (including non-contractual disputes or claims) will be governed by and construed in accordance with the laws of the U.A.E. as applied in the Emirate of Dubai. Each party agrees that the courts of the Emirate of Dubai shall have exclusive jurisdiction to settle any dispute or claim arising out of or in connection with this License Agreement or the Work (including non-contractual disputes or claims).
2. Your rights under this License Agreement will automatically terminate without notice and without the need for a court order if at any point you breach any terms of this License Agreement. In no event will any delay or failure by Bentham Science Publishers in enforcing your compliance with this License Agreement constitute a waiver of any of its rights.
3. You acknowledge that you have read this License Agreement, and agree to be bound by its terms and conditions. To the extent that any other terms and conditions presented on any website of Bentham Science Publishers conflict with, or are inconsistent with, the terms and conditions set out in this License Agreement, you acknowledge that the terms and conditions set out in this License Agreement shall prevail.

**Bentham Science Publishers Ltd.**
Executive Suite Y - 2
PO Box 7917, Saif Zone
Sharjah, U.A.E.
Email: subscriptions@benthamscience.org

**BENTHAM
SCIENCE**

# CONTENTS

This book is dedicated to my son, Jeremy.

# FOREWORD

With all the push toward applications of mathematics where some are at best artificial, it is refreshing to find a text that does not have the pretense of giving any real applications, but rather a book on number theory just for fun. The conception of the book, *Numberama*, could have been conceived by the first real number theorist, P. Fermat, through a bunch of problems (without any thought of applications).

This book is a text about problems in number theory intended to aid teachers from early elementary school to early high school in giving an appreciation of number theory to their students. The text is divided into four parts and an Appendix: Chapter 1 is devoted to several basic problems in number theory which can be appreciated using only elementary arithmetic: addition, subtraction, multiplication, and division. Chapters 2, 3, and 4 are devoted to board games based on the problems in Chapter 1, where each chapter requires successively higher level arithmetic skills to play the games in the chapter. For each of the problems given in Chapter 1, there is a code indicating the level of skills needed to work on at least parts of the problems. Teachers are given plenty of advice as to how to present the material to the students. The Appendix includes excerpts from various student and teacher participants in Dr. Benjamin's Numberama program, which describes both the benefits and the joy they received from participating in this program.

A couple of the problems included are the following, (some going back to antiquity): Finding perfect numbers. Here the students get to use their skills at multiplication and division, as well as being exposed to prime numbers. This problem leads, of course, to open questions such as the existence of infinitely many even perfect numbers and also the existence of at least one odd perfect number. There is also the problem of representing integers as the sum of two squares, a problem which Fermat himself worked on. Again the teachers are given hints as to how to proceed.

Chapters 2, 3, and 4 are devoted to 19 different board games concerning the problems in Chapter 1, which seek to hone the skills of the students, involving many of the properties of numbers given in the first chapter. The text is written well and seems to be accurate. I would certainly recommend it to teachers interested in enriching the mathematics content and honing basic arithmetic skills of the students.

**Chip Snyder, Ph.D.**
Professor of Mathematics,
University of Maine,
Orono,
Maine

# PREFACE

It has been nearly 30 years since I began working on my *Numberama* book. However, in spite of the enormous developments in technology and social media over the past 30 years, the essential theme of my book remains intact. The essential theme is that mathematics can be a stimulating, challenging, and thoroughly enjoyable recreational mental activity to enhance and enrich substantial and creative thinking for children in our school system. As I describe in the Appendix, I have experienced a wide variety of appreciative and enthusiastic responses to my Numberama program over the years, ranging across teacher workshops, teacher education programs, children in gifted programs, children in regular classes, liberal arts college instruction, and family math workshops. I have also effectively utilized my Numberama program at a mental hospital for children, a senior citizen center, as a supplement in my algebra classes, and as an example of creative thinking in my psychology classes. Most recently, I found myself giving a "perfect number lesson" on a napkin at a restaurant at a spiritual development workshop I attended. I had the workshop presenter and some of the participants and staff enraptured, and I realized that Numberama is deeply ingrained in me, wherever I go and whatever I do. I am still a pure mathematician, and I practice what I preach. Doing mathematics for me is refreshing, stimulating, meditative, and enjoyable. I occasionally make use of technology to try to find examples of some of my pure mathematics algebraic number theory results, but this is always very secondary, as the priority is on my "thinking." And this is the exact same philosophy I promote in my Numberama program in regard to the use of technology. Technology is a wonderful tool, but it is essential for the human being to be in control of the technology and not the other way around. Thus finding interesting, surprising, and enticing patterns of numbers, with the assistance of arithmetic calculators when the numbers invariable get very large, is a natural part of my Numberama problems. But what is most important here is the discovery of the patterns themselves, using technology to enhance the discovery.

With this in mind, I am excited to now be offering my *Numberama* book as an e-book through Bentham publications. I welcome feedback from anyone who is using my Numberama problems and games, and I hope that I have succeeded in transmitting the joys of searching for captivating patterns of numbers in my book.

**Elliot Benjamin, Ph.D.**
Instructor of Mathematics at CAL Campus,
Psychology Mentor/Ph.D Committee Chair at Capella University,
Minneapolis,
USA

# ACKNOWLEDGEMENTS

I would like to take this opportunity to give special thanks to a few individuals who truly have made this book possible. First, I am greatly indebted to Dr. Chip Snyder—my mathematical mentor.

I have been working with Dr. Snyder in the pure mathematics discipline of Algebraic Number Theory since I moved to Maine in 1985, having earned my Ph.D. in this mathematical discipline in 1996. We have worked together all these years for essentially one reason—we both enjoyed it—and still do. I learned first-hand the joyful, challenging, frustrating, and transcendental experience of what it means to be a research mathematician. Thus I have been able to practice what I preach.

Next, I must give my heartfelt thanks to my son, Jeremy. As I describe in the Introduction to Chapter 2, it was Jeremy who inspired the games of recreational number theory. It was also my son Jeremy who lived through many of the recreational number theory problems—from ages 7 through 11. He has been wonderfully responsive and patient with his rather unusual father, and I love him dearly.

I must also express my appreciation to a student in my first teacher education in mathematics class at the University of Maine in 1990—Ethel Hill. Ethel thoroughly enjoyed my "special problems." Ethel has also played an enormously important role in making this book a reality—she has done all of the typing! She learned how to put everything on the computer—including the games—in her spare time while working as administrative assistant to the Dean of the College of Education at the University of Maine. She has done this with a marvelous spirit and has encouraged me to persevere in making this book into a vehicle to help in the transformation of mathematics from drudgery to fun. I hope she continues to involve herself in the next stage of marketing this book.

The next individual I give my thanks to is no longer with us—her name is Stephanie Pall. My friend Stephanie was the person who inspired me to find a way to convey to others the joys I have experienced from mathematics. She enabled me to look deeply inside myself, to be who I truly am in my career of teaching mathematics. This occurred in 1988 as I began playing with many of the problems in David Wells' (1986) book.

*The Penguin Dictionary of Curious and Interesting Numbers,* for the purpose of improving my mathematics teaching at Unity College. I subsequently found a number of additional resources that were helpful to me in formulating both my number theory problems and related teaching methods (Adams & Goldstein, 1976; Beiler, 1966; Brown, 1973, 1976, 1983; Dence, 1983; Dewey, 1933; King, 1993; Miller & Heeren, 1961; National Council of Mathematics, 1981, 1984, 1991; Neill, 1960; Postman & U Weingartner, 1969; Rogers, 1969; Walter & Brown, 1971; Walter & Brown, 1977).

Stephanie's untimely death prevented her from seeing where her faith in me has led. But I give tribute to her now; I will forever be indebted to her for the gift she has given me.

I wish to thank all the students in my Finite Math classes at Unity College, my mathematics teacher education classes at the University of Maine, and the teachers and children who participated in my "Mathworks" program in Belfast, Maine, during the 1991−1992 school year.

I also would like to thank my Swanville, Maine, friends who so patiently indulged me in my mathematical entertainments on many a lazy Sunday afternoon in their home—Steve and Kate Webster.

Finally, I give my thanks to Thomas Hathaway Nason from the Word Shop in Orono, Maine, who patiently and effectively put finishing computer touches on the book, a task which turned out to be far more demanding than originally anticipated, and to Kay Retzlaff, my present book consultant who is responsible for the book's new layout and design.

There are many more individuals I am not mentioning who have helped me to form my ideas about both mathematics and mathematics education. I give a final note of thanks to those many unnamed individuals.

# Description of Skill Levels

The following letters will be used to denote the designated math skill levels. All problems and games are followed by the appropriate skill level; problems followed by two or more skill levels imply that they can be used in various degrees of skill complexity. It should be noted that all students can gain value from working on problems from previous skill levels:

1. addition of two-digit numbers
2. general addition and subtraction
3. one-digit multiplication
4. general multiplication
5. multiplication division by 2
6. multiplication division by one-digit number
7. multiplication division in general
8. fractions
9. signed numbers
10. algebra

# INTRODUCTION TO THE BOOK

It is now over 20 years since I wrote the above acknowledgments for this book, as well as the basis of this introduction. However, it is a tribute to the timeless nature of these Recreational Number Theory problems and games that I have designated with the title of *Numberama*, that there is little I feel the need of adding to at this time. My philosophy of "math for fun" has not changed, and I am still collaborating with my ex-Ph.D mathematics mentor Dr. Chip Snyder as we continue to work together, publishing papers in the field of algebraic number theory. I have utilized my Numberama problems and games in diverse educational settings, inclusive of various elementary school classrooms, gifted and talented school programs, developmental mathematics classes at colleges and universities, teacher workshops, and even at a senior retirement home. The past few years I have utilized the Subsets & Circles problem (see Problem #1 in Chapter 1) in my Introductory Psychology classes to illustrate the experience of creative thinking. The results of all my Numberama explorations with both students and teachers have been overwhelmingly positive, and I have received many written descriptions of the benefits that participants have received from their experiences in my Numberama program, a sample of which I have included in the Appendix.

I believe that today, more than ever, it is so very important to not let our children lose (or never experience) the intrinsic joy of doing mathematics. Our technology is so sophisticated that it is all too easy for both our children and ourselves to discontinue our "thinking" and let our computer gadgets "think" for us. But there is an inherent potential joy in thinking, and I am thankful that I continue to experience this inherent joy of mathematical thinking in my pure mathematics field of algebraic number theory. And it continues to be part of my mission in life to convey the inherent joy of mathematical thinking to children in the context of Numberama Recreational Number Theory problems and games in the school system, and to people of all ages, through my *Numberama* book.

As a child I enjoyed adding numbers in my head. People were amazed at how quickly I could do so without using pencil or paper. Throughout school I enjoyed math and, as a result, I was good at it. It was no surprise to anyone when I decided to become a math teacher; however, I soon realized that the intrinsic rewards I received from studying mathematics were by no means a common experience for other students. After teaching elementary and high school, college, and in various adult education programs, I came to the conclusion that the vast majority of our population has a very limited perspective of what mathematics is truly all about.

Mathematics can certainly be an extremely pragmatic science, chock full of useful applications in virtually every field studied; however, there is another side to mathematics. Pure mathematics can be described as an art form, in the same way music, art, dance and theater are arts. Nearly every professor of mathematics knows this deep down in his/her heart. Mathematics is truth and beauty within the spirit of the mind. The natural process of thinking is inherently pleasurable. Pressures, grades, competition, *etc.*, can destroy this potential intrinsic pleasure. What I refer to as a "natural dimension of mathematics" is doing math for the pure enjoyment of learning and discovering. Math can be fun.

This book attempts to impart the enjoyment of mathematics to the children in our schools, whether these schools are at home or part of a public or private system. The branch of mathematics that literally plays with numbers is known as number theory. Topics in number theory range from the highly theoretical, employing deep layers of abstract mathematical proof, to questions about numbers that any school child learning arithmetic can understand. These questions are enticing, adventuresome, challenging, and most important of all—fun.

I call this form of number theory, recreational number theory. Most of the problems described in Chapter 1 in this book can be worked by children who know how to add, subtract, multiply, and divide. A number of the problems do not even require division; some of the problems only require addition. There are also problems for children first learning fractions, and in many of the problems I have given suggestions on how they can be formulated into algebra problems for older students, in junior and senior high school. For each problem, the exact prerequisite skills are indicated. The general format is described at the end of the Table of Contents. The problems I have chosen to describe are by no means exhaustive. An examination of the bibliography I have included will give the interested reader some supplementary material. There is a place in our schools for "math for fun" problems. The earlier such problems are introduced, the easier it will be for a child to learn the basics of arithmetic. Working on these problems requires a lot of practice in nearly all of the arithmetic skills that are now being taught in the elementary schools. But the practice and drill are made fun through the discovery of patterns, formulas, unusual numbers, *etc*. The approach I am recommending is very much like playing a game.

Chapters 2, 3, and 4 consist of a series of 19 games based upon the ideas from recreational number theory introduced in Chapter 1, with each chapter requiring successively higher arithmetic skills for children to play the games included in the chapter. These games hones the skills of the students, involving many of the properties of numbers given in the first chapter. Once again, the games are by no means exhaustive, but merely serve as a rough.

Mathematics can certainly be an extremely pragmatic science, chock full of useful applications in virtually every model of how many math ideas can be made into games where children are joyfully practicing their arithmetic skills while playing the game. The prerequisite skills necessary to play the games are listed for each game, in the same format described for the problems in Chapter 1. These games serve to reinforce ideas encountered in the problems. Although a major emphasis of recreational number theory is the elementary school, this is by no means the only place where it can be used. I have purposely included many generalizations to algebraic formulas in order to make the point that recreational number theory can be used throughout the school years.

Junior high and high school students can be taught to use their newly acquired algebra skills to generate their own algebraic formulas that describe experimental facts about numbers that they have gathered together. This approach to teaching algebra is a radical change from the often tedious, monotonous, and overly pragmatic way that algebra is generally taught in our school system. I am by no means recommending that all of the traditional material in arithmetic or algebra be deleted from our schools; rather I am advocating an exciting new tool and method of education that can be used to help our children learn many of these skills. The key word is "balance." There is a place for lecture, a place for tradition, and also a place for process, adventure, and discovery.

Another challenge is to successfully use the discovery approach of recreational number theory with college students in the area known as developmental mathematics, which is little more than arithmetic and algebra for college students and adults going back to school. Community colleges and continuing education departments are teaching more of arithmetic and algebra to their students than any other kind of math. For much of my career as a mathematics professor, this was my own specialized field, and math anxiety, resistance, built-up failures, *etc.*, are painfully high in this student population. To enable these students to view mathematics as a pleasurable pastime is indeed challenging.

This is the challenge this book is intended to meet. I have seen extremely dramatic results with students who hated math all of their lives. The prospect that they could now play with numbers for the purpose of making joyful discoveries was a welcome change of pace for them; however, the results were best when I was able to use the discovery approach exclusively without having to worry about required topics, exams, and follow-up courses. I realize this is not the typical situation our students are in, and throughout my mathematics college teaching years.

I continued to search for an effective way of balancing the old and the new; *i.e.*, to incorporate the ideas and processes of recreational number theory within the traditional format of our developmental mathematics courses.

I hope that you will find value in the following problems and games, whether you are a math teacher, prospective math teacher, math student, parent, or interested reader in general. I welcome any feedback you have, and look forward to hearing from you.

## INTRODUCTION TO THE GAMES

When my son Jeremy was 7 years old, he made me a little math game for Father's Day. He seemed to think that it would be fun to play games based upon some of the math ideas I had been trying out on him, and he made me a cute little precursor of the Syracuse Algorithm Game. I took my son's idea seriously, as you can see from the games in Chapters 2, 3, and 4 of this book. The 19 games I am including are only a sample of the kinds of games you can make out of the Recreational Number Theory problems introduced in Chapter I. Many of the games described in Chapters 2, 3, and 4 have been played by elementary school teachers in some of my Numberama teacher workshops. Teachers generally found them to be a productive, fun-loving way of helping children learn their arithmetic skills. They also had some excellent suggestions in terms of modifying various aspects of the games (see the section below for their suggestions). However, please keep in mind that my purpose in describing these games is only to offer you a basic framework. It is my hope that you will develop the game ideas for yourself, according to your own unique needs, interests, imagination, and artistic capabilities. Lastly, it is important to keep in mind that the theme for all of the games in Chapters 2, 3, and 4 and all of the problems in Chapter 1, is that the children should be having fun while they are learning mathematics. The game equipment that I have used are gameboards, dice, number cards, play money—in all denominations from $1 to $1,000, Lego® pieces for the players, and game rules. Chapters 2, 3, and 4 are divided according to the required arithmetic skill levels for children to play these games.

# GAME IDEAS FROM TEACHERS AT NUMBERAMA WORKSHOPS

Some ideas that came out of my Numberama teacher workshops in regard to the games are as follows:

1. Form teams of two or more children.
2. Put a time limit on how long a child can take to give an answer.
3. Use an attachment to the gameboard instead of separate number cards.
4. Give a child at least some money even when an answer is incorrect.
5. Make the games colorful and artistically attractive.
6. Include paper with numbered items for a child to keep a record of.
7. Instruct all children to work on the problem while a child has a turn.
8. Encourage children to make exchanges of money in the game.

I find all of these ideas to be excellent suggestions, and I'm sure you will have more of your own as you read through Chapters 2, 3, and 4. My own suggestion is to play the game after the children have had a chance to explore the ideas that the game is based upon. In other words, I believe the games will be most effective when the children have already experienced the discovery processes described in Chapter 1.

# Recreational Number Theory Problems

**Abstract:** Chapter 1 comprises the nuts and bolts of Numberama, as it includes all the problems that I have included in Recreational Number Theory as part of my Numberama program. Each problem has the designated skill level required, and the problems begin with the Subsets and Circle problem, which utilizes only addition and subtraction, though a knowledge of basic algebra would be required for middle school or high school students to understand the algebraic formulations of this problem. Early in the sequence of problems I go to the Perfect Numbers problem, which admittedly is my favorite problem, and requires a knowledge of multiplication and division, along with once again a knowledge of basic algebra for the same purpose as the Subsets and Circle problem. The learning and teaching strategies that I have included in the Perfect Numbers problem are especially effective in awakening students to the mystery, surprise element, and beauty inherent in our number system, as well as developing an understanding of mathematics as an exciting open-ended field of study with unknown problems that can be worked on with high level computers. Later on in my sequence of problems, one encounters the Sums of Squares problem, which goes back to Fermat, one of the original founders of Number Theory, and once again is an excellent way to develop an appreciation in students of the beauty and mystery involved in our number system. Taken as a whole, these problems from Recreational Number Theory that I have chosen to utilize in my Numberama program can serve as the "magic" needed to demonstrate the inherent joy of mathematics to our students in the school system—and to people of all ages.

**Keywords:** Abundant numbers, Circles, Clock arithmetic, Deficient numbers, Farey fractions, Fibonacci numbers, Goldbach conjecture, Kaprikar numbers, Linear diophantine equations, Magic squares, Multiplicative persistence, Pascal's triangle, Perfect numbers, Powerful numbers, Prime numbers, Semi-Prime numbers, Subsets, Sums of squares, Triangular numbers, Weird numbers.

## 1. SUBSETS AND CIRCLES (A, B, J)

This first problem is appropriate for all ages, kindergarten through college. As with most of the problems, it can be done either from the elementary level of arithmetic or from a more advanced perspective of first-year algebra. I recommend a series of three stages in this problem, as follows:

Elliot Benjamin

## Stage 1

Define a set as a collection of objects: a set of chairs, set of people, set of numbers, *etc.* Describe the set of numbers as {1,2}, {2,4,6}, {2,5}, {6}, *etc.* Define a subset as part of a set; thus {1,2} is a subset of {1,2,3}; {2} is a subset of {2,5}, but {2,4} is not a subset of {2,5}. Include both the whole set and the empty set Ø—consisting of no elements—as subsets of every set. Thus, all the subsets of {1,2,3} are {1}, {2}, {3}, {1,2}, {1,3}, {2,3}, {1,2,3}, and Ø. Have the children make a table like the following (Table **1**):

Table 1. Stage 1a of subsets & circles problem.

| Number of Numbers | Number of Subsets |
| --- | --- |
| 3 | 8 |

Depending upon the age range you are dealing with, work out the subsets with the children for sets up to three or four numbers, and have them do the next set of numbers on their own. You will find the pattern doubles in the following way (Table **2**):

Table 2. Stage 1b of subsets & circles problem.

| Number of Numbers | Number of Subsets |
| --- | --- |
| 0 | 1 |
| 1 | 2 |
| 2 | 4 |
| 3 | 8 |
| 4 | 16 |
| 5 | 32 |

Ask the children to say what the pattern is and guess how many subsets they think would be in a six-element set. For grades 4 through 6, ask the children how many subsets would be in a 10-element set, a 12-element set, *etc.* Ask algebra students to determine the algebraic formula that describes the number of subsets in any set of **n** elements. Hint that it has something to do with exponents. The answer is that for any number n, the number of subsets in a set of **n** numbers is $2^n$, as can easily be experimentally verified by Table **2**.

## Stage 2

Now that the basic idea of experimenting with numbers and finding a pattern has been established, the children are ready for a true adventure in mathematics.

Draw a circle with two points—spaced equally apart—on it and connect the two points with a line—in the following way (Fig. **1**):

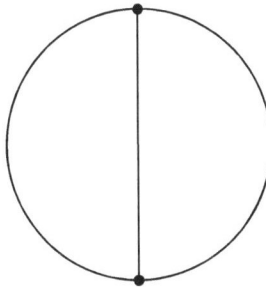

**Fig. (1).** Stage 2A of Subsets & Circles Problem.

Make a chart like the following:

| Number of Points | Number of Regions |
|---|---|
| | |

The children will observe that when there are two points on the circle, the circle is divided into two regions. Continue this process until five points on the circle are reached—always spacing the points equally apart. The circle will look like the following (Fig. **2**):

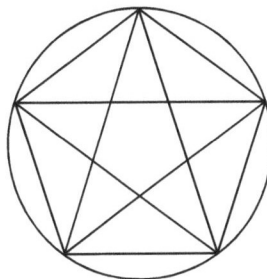

**Fig. (2).** Stage 2B of Subsets & Circles Problem.

The children will undoubtedly observe that the number of regions is doubling as the number of points is increased, as can be seen from the following table (Table **3**):

**Table 3. Stage 2 of subsets & circles problem.**

| Number of Points | Number of Regions |
|:---:|:---:|
| 2 | 2 |
| 3 | 4 |
| 4 | 8 |
| 5 | 16 |

Ask the children how many regions they estimate for a six point circle and have them draw their own circle—of course expecting to get 32 regions. To their astonishment, they will not get 32 regions. Depending on how accurately they space their points, they will get 30 or 31 regions. After proving to themselves, by doing it over once or twice, that the correct answer is either 30 or 31 (30 if drawn with equal spaces between points), but certainly not 32, ask them if they think this is just a coincidence, or is there a reason for the pattern, 2, 4, 8, 16, 30 or 31? They are now ready for the final part of the problem.

**Stage 3**

Draw the two-point circle again, but this time include in your table the number of "edges," defined as any connection between two points on the circle, including lines and curves (Fig. **3**):

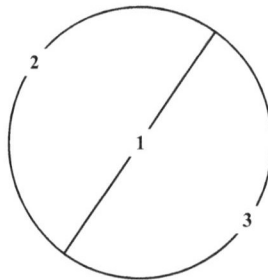

**Fig. (3).** Stage 3A of Subsets & Circles Problem.

Continuing the process again for a three-point circle, the following table will be obtained (Table **4**):

**Table 4. Stage 3a of subsets & circles problem.**

| Number of Points | Number of Regions | Number of Edges |
|:---:|:---:|:---:|
| 2 | 2 | 3 |

*(Table 4) contd.....*

|  |  |  |
|---|---|---|
| 3 | 4 | |

For a four-point circle, a new definition of point must be made. "Point" is now taken to mean both points on the outside of the circle and on the inside of the circle. Thus the original four-point circle now has a total of five\points associated to it (Fig. **4**):

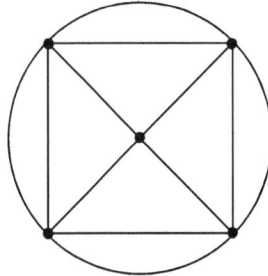

**Fig. (4).** Stage 3B of Subsets & Circles Problem.

Have the children continue the process until the original five-point circle is reached. The table will look like the following (Table **5**):

Table 5. Stage 3b of subsets & circles problem.

| Number of Points | Number of Regions | Number of Edges |
|---|---|---|
| 2 | 2 | 3 |
| 3 | 4 | 6 |
| 5 | 8 | 12 |
| 10 | 16 | 25 |

Ask the children if they observe any pattern. Give them a hint that it has something to do with the relationship amongst the points, regions, and edges. Ask them to fill in the remaining data for the original six-point circle. Depending upon whether they used 30 or 31 regions they should obtain the following data (Table **6**):

Table 6. Stage 3c of subsets & circles problem.

| Number of Points | Number of Regions | Number of Edges |
|---|---|---|
| 19 | 30 | 48 |
| 21 | 31 | 51 |

The pattern should be clear to at least a few students that the number of points plus the number of regions is always one more than the number of edges; an equivalent formulation is that the number of points plus the number of regions minus the number of edges always equals one. With this discovery a whole new perspective on mathematics has been demonstrated: one of excitement, adventure, mystery, and profound harmony. For algebra students, stress the algebraic relationship amongst points, regions, and edges: $P + R = E + 1$, known as "Euler's Formula". Discuss how using a little elementary algebraic manipulation, the formula can be described equivalently as:

$$P + R - E = 1, E - R = P - 1, E - P = R - 1, etc.$$

## 2. MULTIPLICATIVE PERSISTENCE (C)

This problem is nice and easy, perfect for children first learning how to multiply. Ask your students to give you a two-digit number, 36 for example, and tell them to multiply the digits together: $3 \times 6 = 18$. Then tell them to multiply the digits together again, continuing this process until a single digit number is reached: the sequence beginning with 36 therefore is 36-18-8, and we say that the multiplicative persistence of 36 is 2; this simply means that it took two steps to reach a single-digit number.

Ask the children to give you some other two-digit numbers until you get numbers that have multiplicative persistences of 1, 2, and 3. For example, 24 has multiplicative persistence of 1; 24-8, and 97 has multiplicative persistence of 3; 97-63-18-8.

Ask the children if they think they can find a two-digit number with multiplicative persistence of 4. There is one—and only one. Hopefully none of the numbers given to you is this unique number. Otherwise you can give your class extra recess or lunch time. The unique two-digit number that has multiplicative persistence of 4 happens to be 77: 77-49-36-18-8.

## 3. SYRACUSE ALGORITHM (E, J)

This problem is tricky. The algorithm is as follows: pick a number—one- or two-digit—and if the number is odd, multiply it by 3 and add 1 to it; if it is even, take half of it (algebraically this can be conveniently stated in function language as $f(x) = 3x + 1$ if x is odd; $f(x) = x/2$ if x is even). Then continue the process. For example, if you were to choose 13 the sequence would be 13-40-20-10-5-16-8-4-2-1.

Once you get 1 the process will repeat interminably as 1-4-2-1-4-2-1, *etc.*, so we will say that the sequence ends with 1.

Notice that it took nine steps to get from 13 to 1. Ask your students to pick another number: perhaps they'll choose 19. The sequence is: 19-58-29-88-44- 22-11-34-17-52-26-13-40-20-10-5-16-8-4-2-1; there were 20 steps.

What is so tough about this problem? Ask your students to do the Syracuse Algorithm for 27—and after half an hour you might want to make sure there are no eggs in the room. For 27 is one of a few two-digit numbers that takes over a hundred steps; it takes 111 steps!

This provides good practice in multiplying by 3, dividing by 2, and especially in persistence and accuracy. It also demonstrates an unsolved problem in mathematics, as it is not known whether or not applying the Syracuse Algorithm makes all numbers eventually reach 1. The first billion numbers have been tested using high speed computers and they all do eventually reach 1, but mathematicians have not been able to prove that this process works for all possible numbers; neither have they ever found a number for which the Syracuse Algorithm does not work.

Portraying mathematical problems whose solutions are unknown to children who can understand and actually even work on these problems is a marvelous way of inculcating the "prime" values of exploration, adventure, and discovery in mathematics to children of school age.

## 4. MAGIC SQUARES (A)

We come now to an age-old puzzle type of problem, enjoyed by none other than Benjamin Franklin while sitting in congress in the Pennsylvania Assembly. The following diagram is a 4x4 magic square (Table **7**):

**Table 7. First table of magic squares problem.**

| 16 | 3 | 2 | 13 |
|----|----|----|----|
| 5 | 10 | 11 | 8 |
| 9 | 6 | 7 | 12 |
| 4 | 15 | 14 | 1 |

Notice that the sum of each of the rows, each of the columns, and each of the diagonals happens to be 34. 34 is called the "magic number" of this square. In this particular square each of the numbers from 1 to 16 is used exactly once. For a 5x5 magic square look at (Table **8**):

**Table 8. Second table of magic squares problem.**

| 15 | 18 | 21 | 4 | 7 |
|----|----|----|----|----|
| 24 | 2 | 10 | 13 | 16 |
| 8 | 11 | 19 | 22 | 5 |
| 17 | 25 | 3 | 6 | 14 |
| 1 | 9 | 12 | 20 | 23 |

Notice that in this 5x5 square, each of the numbers from 1 to 25 is used exactly once and the magic number is 65. A few amazing 3x3 magic squares, using only prime numbers (numbers with no divisors other than the number itself and 1) are (Table **9**):

**Table 9. Third, fourth, and fifth tables of magic squares problem.**

| 479 | 71 | 257 |
|----|----|----|
| 47 | 269 | 491 |
| 281 | 467 | 59 |
| 389 | 227 | 191 |
| 71 | 269 | 467 |
| 347 | 311 | 149 |
| 401 | 257 | 149 |
| 17 | 269 | 521 |
| 389 | 281 | 137 |

Come up with a 3x3 magic square, using each of the numbers from 1 to 9 exactly once, with a magic number of 15. Is it hard or easy?

## 5. PERFECT NUMBERS (F, G, J)

The next topic is excellent practice for students first learning division. However, it quickly goes way beyond its pragmatic use of division practice to an appreciation of some of the most notorious unsolved problems in the current world of mathematics. As the topic of perfect numbers has many ramifications, we will take this unusual topic in five stages (see Appendix 3 for my essay on using perfect numbers to work with a gifted student).

## Stage 1

Define a perfect number as a number such that all the numbers that divide into it evenly—not including the number itself—add up to the original number. For example, all the numbers that divide into 8 evenly (called proper divisors of 8) are 1, 2 and 4. They add up to 7 and therefore 8 is not a perfect number. However, 6 has proper divisors 1, 2, and 3: they add up to 6, and 6 is the first (smallest) perfect number.

Ask your students to determine what the second perfect number is—give them a hint that it is less than either 30 or 50—depending upon how much one-digit division practice you would like them to have. It shouldn't take too long before a child comes up with 28 to be the second perfect number: $28 = 1 + 2 + 4 + 7 + 14$.

## Stage 2

Now the fun starts. Inform your students that the third perfect number is rather large; however, it turns out that there is an interesting pattern for perfect numbers.

Describe to them how the first two perfect numbers, 6 and 28, can be written as 6 = 2 x 3, 28 = 4 x 7, and ask them if they see any possible pattern to guess what the third perfect number is.

Take their guesses and have them check to see if their candidates for perfect numbers are correct. A pattern might be 2 x 3, 4 x 7, and 8 x 11 = 88. Another possible pattern is 2 x 3, 4 x 7, 8 x 21 = 168; or 2 x 3, 4 x 7, 6 x 21 = 126; or 2 x 3, 4 x 7, 16 x 11 = 176, *etc.*

As the numbers to be tested are getting larger, explain to the children a few helpful divisibility rules that will shorten the number of required divisions:

1. A number is divisible by 5 if and only if it ends in either 5 or 0;
2. A number is divisible by 2 if and only if it ends in either 2, 4, 6, 8, or 0;
3. A number is divisible by 3 (respectively 9) if and only if the sum of all the digits of the number is divisible by 3 (respectively 9)
4. A number is divisible by 6 if and only if it is divisible by both 2 and 3;
5. When testing for possible divisors of a number, test only as high as one less than a number whose square is higher than the original number.

For example, if testing all numbers to see if they divide evenly into 150, do not test any number higher than 12, since $13^2 = 169$.

After a number of candidates have not generated the third perfect number, you might offer a possible pattern: 2 x 3 = 6, 4 x 7 = 28 and 8 x 15 = 120. The reasoning for this pattern is that $7 = 3 \times 2 + 1$, and $15 = 7 \times 2 + 1$. This number, 120, will also fail to be perfect.

However, ask them what 3 and 7, the second factors of 6 and 28, have in common with each other that 15, the second factor of 120, does not.

The answer will come that 3 and 7 have no divisors other than themselves and 1, whereas 15 has both 3 and 5 as divisors. Thus the crucial notion of "prime number" has finally been acknowledged.

Ask the children to continue the pattern one more time: 2 x 3, 4 x 7, 8 x 15, and 16 x 31 = 496. It should be clear to everyone that 31 is a prime number; ask them to verify that 496 is indeed the third perfect number.

**Stage 3**

At this point it should be understood what needs to be done in order to find the fourth perfect number.

(A useful tricky false pattern that generates the first three perfect numbers but breaks down for the fourth perfect number is: 2 x (2x2 − 1) = 6, 4 x (2x4 − 1) = 28, 16 x (2 x 16 − 1) = 496).

Continue the process, until the second factor is once again a prime number: 2 x 3, 4 x 7, 8 x 15, 16 x 31, 32 x 63, and finally 64 x 127 = 8,128.

Ask the children to first show that 127 is a prime number and that 8,128 is a perfect number. Doing the divisions directly would require checking all numbers through 90, since $91^2 = 8,281$, which is the smallest square greater than 8,128. However, they can use the following very useful shortcut which utilizes the "Fundamental Theorem of Arithmetic".

Demonstrate how any number can be factored uniquely into a product of prime numbers with exponents. For example, $100 = 5^2$ x $2^2$, $90 = 2$ x $3^2$ x 5, $16 = 2^4$, *etc.*

Therefore, in determining divisibility it is sufficient to just look at the primes in a number's prime factorization. For example, if a number is not divisible by each of the primes: 2, 3, and 5, then it cannot be divisible by 90, since $90 = 2$ x $3^2$ x 5.

This will eliminate many divisions when dealing with large numbers. In fact, this procedure leads to the fact that no even number can divide into an odd number, since 2 is in the prime factorization of every even number and a number is divisible by 2 if and only if the number ends in 2, 4, 6, 8, or 0; *i.e.*, if the number is even. At any rate, after a number of divisions it should be verified that 8,128 is indeed the fourth perfect number, and that our pattern/formula does indeed seem to consistently yield perfect numbers.

## Stage 4

The last perfect number students ought to obtain is the fifth one: 33,550,336. The exercise is appropriate and beneficial for children who are practicing long division. The way to do it is to continue the pattern, carefully checking to see if the second factors are prime numbers.

The second factor will be (continuing from 127 for the fourth perfect number): 255, 511, 1023, 2047, 4095, and finally 8,191—which is the next prime number in the sequence.

It is a challenging exercise to verify that 33,550,336 is indeed perfect. Once these five perfect numbers are obtained (or four are obtained in the case of younger children), you might ask your students what all of these perfect numbers seem to have in common.

Someone will probably observe that the numbers all end in either 6 or 8, actually alternating from 6 to 28. You can generate the hypothesis that all perfect numbers end in either 6 or 8 in alternating fashion. Then you can inform the children that this has been shown to not be the case; *i.e.*, the sixth perfect number ends in 6, not 28 (8,589,869,056).

Ask for another observation, and someone will most likely say that all the perfect numbers are even.

Finally, you come to another demonstration of an unsolved problem in mathematics: does an odd perfect number exist? You can tell the children that so far, using high speed computers, 33 perfect numbers have been discovered—and they are all even (and end in 6 or 28, which happens to be a consequence of a perfect number being even). Nobody knows if an odd perfect number exists.

You might also ask how many perfect numbers there are? Talk about the idea of there being infinitely many counting numbers, even numbers, odd numbers, multiples of a number, and prime numbers. Then state a second unsolved problem about perfect numbers: does there exist infinitely many perfect numbers?

Lastly, ask algebra students to describe the algebraic formula for obtaining even perfect numbers. Give them the hint that the formula is related to the formula for the number of subsets—$2^n$—of a set with n elements. With some constructive leads, they should arrive at the formula: $2^n \times (2^{n+1} - 1)$ where $2^{n+1} - 1$ is a prime.

## Stage 5

Later, after the theory of perfect numbers has been digested and largely forgotten, you might bring up the subject again with the following amusing proposition: tell them the next pattern does not work for 6, the first perfect number, but note that the second perfect number $28 = 1^3 + 3^3 = 1 + 27$, and the third perfect number $496 = 1^3 + 3^3 + 5^3 + 7^3 = 1 + 27 + 125 + 343$.

Ask the students to see if this pattern of adding successive cubes of odd numbers will yield the fourth perfect number. They should be able to verify that $8,128 = 1^3 + 3^3 + 5^3 + 7^3 + 9^3 + 11^3 + 13^3 + 15^3$.

Next, for excellent practice in multiplication, exponents, and record-keeping, have the students demonstrate that the pattern even works for the fifth perfect number: 33,550,336. They will need to go up until the cube of 127 to get it to work. You can inform them that all known perfect numbers, other than 6, can be written as the sum of consecutive odd cubes. With this you can end the subject matter of perfect numbers.

## 6. SEMI-PRIME NUMBERS (F)

An amusing little change of pace can be done using what is known as semi-prime numbers. Semi-prime numbers are numbers that have exactly four factors. For example, 15 is a semi-prime number because the factors of 15 are 1, 15, 3, and 5. Twelve is not a semi-prime number because it has 6 factors: 1, 12, 2, 6, 3, and 4; neither is 9 a semi-prime number since it has three factors: 1, 9, and 3.

An informal way of thinking about it is that semi-prime numbers are non-square numbers that are almost prime. Semi-prime numbers are plentiful enough, but the problem is to come up with consecutive semi-prime numbers. For example, 14 and 15 are two consecutive semi-prime numbers. The problem for the children is to come up with three consecutive two-digit semi-prime numbers. After a little trial and error, most of your students should find the three consecutive semi-prime numbers: 33, 34, and 35.

You might ask them if they can also find the second set of three consecutive two-digit semi-prime numbers, or perhaps ask them for two sets to begin with, depending upon how much one-digit division you would like them to practice. The second set is 85, 86, and 87.

## 7. ABUNDANT AND DEFICIENT NUMBERS (G)

We now briefly return to the fascinating topic of perfect numbers. Recall that a number is perfect when all of the proper divisors of the number add up to the number itself. Recall also that there were very few perfect numbers.

If a number is not perfect, then obviously all of the proper divisors of the number must either add up to less than the number or greater than the number. If the sum of the proper divisors of a number is less than the number itself, we call the number "deficient". If the sum of the proper divisors of a number is greater than the number itself, we call the number "abundant".

For example, 8 is deficient since the proper divisors of 8 are 1, 2 and 4, and $1 + 2 + 4 = 7$; 12 is abundant since its proper divisors are 1, 2, 3, 4, and 6, and $1 + 2 + 3 + 4 + 6 = 16$.

Now ask the children to begin testing numbers for abundancy and deficiency. Have them compare this in a table for two-digit numbers. What they should find is that all of the abundant numbers they have found are even.

This suggests the following question: does an odd abundant number exist? To answer the question they will now have to deal with three-digit numbers (*i.e.*, numbers between 100 and 1,000).

This is excellent long division practice. You might tell them in advance that the answer is "yes," (*i.e.*, there is an odd abundant number, and there is one with three digits—but there is only one with three digits).

The children's job is to find the number. The hint is to use numbers for which lots of small odd numbers go into evenly. Point out to the students that even numbers do not divide into odd numbers without leaving a remainder. Thus they might try the number $3 \times 5 \times 7 = 105$. The proper divisors of 105 are 1, 3, 5, 35, 21, 7, and 15 and they add up to 87; not quite abundant, but getting there. Perhaps $3 \times 5 \times 9$ would work, or $5 \times 7 \times 9$, or $3 \times 7 \times 11$, *etc*. What is the correct answer? Is it $3 \times 5 \times 7 \times 9 = 945$?

## 8. WEIRD NUMBERS (F)

Just as there are semi-prime numbers, there are also semi-perfect numbers. By definition, a semi-perfect number is an abundant number, "some" of whose proper divisors add up to the number.

For example, 12 is a semi-perfect number since some of the divisors of 12 are 2, 4 and 6, and the sum of these three numbers is 12. Similarly 30 is a semi-perfect number since some of the divisors of 30 are 5, 10, and 15, and the sum of these three numbers is 30. Obviously a deficient number cannot be semi-perfect.

The problem for this topic is to come up with a two-digit abundant number that is not semi-perfect, (*i.e.*, a two-digit weird number). It turns out that there is only one two-digit number with this property—only one "weird" two-digit number (and, incidentally, there is only one "weird" three-digit number). Have fun!

## 9. SUMS OF SQUARES (F, G, J)

We now come to a problem that is a cornerstone of number theory. Sums of squares simply means writing a number as the sum of two perfect squares: for example, $17 = 16 + 1 = 4^2 + 1^2$, $45 = 36 + 9 = 6^2 + 3^2$, *etc.*

The problem can be stated algebraically as $n = x^2 + y^2$ where n, x, and y are all positive integers, (*i.e.*, whole numbers).

Solving equations for integers is known as solving "Diophantine equations," which represents the most active area of research being conducted in the field of number theory today. We will therefore divide this far-reaching topic into five stages.

### Stage 1

Begin by asking what prime numbers can be represented as the sum of two squares. For example, $13 = 3^2 + 2^2$ but 19 cannot be written as the sum of two squares.

For a simple check on 19, take $1^2 + 18$, $2^2 + 15$, $3^2 + 10$, $4^2 + 3$, and notice that $5^2$ is too big, demonstrating clearly that 19 cannot be written as the sum of two squares, the reason being that 18, 15, 10, and 3 are not perfect squares and these are the only possibilities for the other square.

Have the children compose a table which looks like the following (Table **10**):

Table 10. Stage 1a of sums of squares problem.

| Yes | No |
|---|---|
| $1^2 + 1^2 = 2$ | 3 |
| $2^2 + 1^2 = 5$ | 7 |
| $3^2 + 2^2 = 13$ | 11 |
| $4^2 + 1^2 = 17$ | 19 |
| $5^2 + 2^2 = 29$ | 23 |

Ask them to try to guess where the next prime number—31—will go. Have them search for a pattern that determines whether or not a prime number can be written as the sum of two squares. Ask them to continue the table for all primes less than 100.

Most likely no one will discover the pattern, including yourself. A hint—which will certainly lead to the solution of the problem—is to find the difference between all successive primes in both the Yes and No columns, not including the number 2. Thus, your table should now look like Table **11**:

**Table 11. Stage 1b of sums of squares problem.**

| | Yes | No |
|---|---|---|
| | 2 | 3 |
| | | ) 4 |
| | 5 | 7 |
| 8 ( | | ) 4 |
| | 13 | 11 |
| 4 ( | | ) 8 |
| | 17 | 19 |
| 8 ( | | ) 4 |
| | 29 | 23 |
| 8 ( | | ) 8 |
| | 37 | 31 |
| 4 ( | | ) 12 |
| | 41 | 43 |

Notice that the differences are all multiples of 4. Ask the students where the prime 101 should go, and it may be fairly evident to them that 101 will go into the Yes column because $101 - 41 = 60$ is a multiple of 4, while $101 - 43 = 58$ is not a multiple of 4; indeed $101 = 10^2 + 1^2$. Similarly 103 goes in the No column because $103 - 43 = 60$ is a multiple of 4 while $103 - 41 = 62$ is not a multiple of 4.

A more mathematical way of stating this result is to first note that every number when divisible by 4 either goes in evenly or leaves a remainder of 1, 2, or 3. Odd numbers must leave remainders of 1 or 3, all prime numbers other than 2 are odd, and therefore all prime numbers, other than 2, when divided by 4 leave a remainder of 1 or 3. If the remainder is 1, it can be written as the sum of two squares; if the remainder is 3 it cannot be written as the sum of two squares.

Algebraically this can be simply stated as all prime number of the form $4k + 1$, where k is a positive integer, can be written as the sum of two squares, and all prime numbers of the form $4k - 1$ (or equivalently $4k + 3$) cannot be written as the sum of two squares. We have thus completely solved the Diophantine equation $n = x^2 + y^2$ when $n$ is a prime number.

## Stage 2

Now that we so clearly know when we can write prime numbers as the sum of two squares, we will ask the analogous question for any number; *i.e.*, when are we able to write any positive integer as the sum of two squares?

Algebraically speaking, we are asking when we can solve the Diophantine equation $b = x^2 + y^2$ for any positive integer b. The solution to this problem relates directly to the previous problem for prime numbers, having to do with the "Fundamental Theorem of Arithmetic"—which says that any number can be factored uniquely as a product of prime numbers with exponents: for example, $100 = 5^2 \times 2^2$, $80 = 2^4 \times 5$, $210 = 2 \times 3 \times 5 \times 7$, *etc.*

This can be described algebraically as $b = p_1^{e(1)} \cdot p_2^{e(2)} \cdot p_3^{e(3)} \ldots p_n^{e(n)}$, where $p_1$, $p_2$, ... $p_n$ are the unique primes and $e_1$, $e_2$, $e_3$ ... $e_n$ are the unique exponents.

Of course the order of the prime numbers can be changed; for example, $210 = 2 \times 3 \times 5 \times 7 = 5 \times 7 \times 2 \times 3 = 3 \times 7 \times 2 \times 5$, *etc.* However, before tackling the general problem of writing any number as the sum of two squares, you might ask the children to find a two-digit number that can be written as the sum of two squares in two different ways, without using 0 as one of the squares.

Following this requirement, do not count $25 = 5^2 + 0^2$ and $25 = 3^2 + 4^2$ as two different ways. Tell them that there are three two-digit numbers that have this property, and it should not take them too long to come up with at least one of these numbers: $50 = 7^2 + 1^2$ and $50 = 5^2 + 5^2$, $65 = 8^2 + 1^2$ and $65 = 7^2 + 4^2$, and 85

$= 9^2 + 2^2$ and $85 = 7^2 + 6^2$.

Ask them to find three numbers between 100 and 200 that have this same property, and, after some trial and error, they might come up with $130 = 11^2 + 3^2$ and $130 = 9^2 + 7^2$, $170 = 13^2 + 1^2$ and $170 = 11^2 + 7^2$, and $185 = 13^2 + 4^2$ and $185 = 11^2 + 8^2$.

## Stage 3

The children are now ready to solve the general Diophantine equation $b = x^2 + y^2$, or more plainly speaking, to write any number as the sum of two squares. Tell them it is related to the first problem they worked on with prime numbers. Compose with them the following type of table (Table **12**):

**Table 12. Stage 3 of sums of squares problem.**

| Yes | | No | |
|---|---|---|---|
| Prime Factors | Number | Prime Factors | Number |
| $13 \times 5^2$ | $8^2 + 1^2 = 65$ | $7 \times 3$ | 21 |
| $3^2 \times 21$ | $3^2 + 3^2 = 18$ | $3^2 \times 2$ | 54 |
| $2^2 \times 5^2$ | $6^2 + 8^2 = 100$ | $5^2 \times 3$ | 75 |
| $7^2 \times 21$ | $7^2 + 7^2 = 98$ | $7 \times 2$ | 14 |
| $3^4 \times 21$ | $9^2 + 0^2 = 81$ | $3^3$ | 27 |

It will be evident to the children that a number for which all the odd primes in its factorization with remainder 1—when divided by 4—is always a Yes. However, it is not true that a number with only primes in its factorization that have remainder 3 —when divisible by 4—is always a No, as $18 = 3^2 \times 2$ is a Yes; $98 = 7^2 \times 2$ is a Yes; $81 = 3^4$ is a Yes, *etc.*

Give the students the hint that it has something to do with the exponents. Some children will realize that numbers with only primes in their factorization of the form $4k + 1$, or $4k - 1$ with an **even** exponent, result in Yes's, and numbers that have any primes of the form $4k - 1$ with an **odd** exponent in their factorization result in No's.

In other words, any prime in a number's prime factorization which leaves a remainder of 3—when divisible by 4—and is taken to an odd exponent (including 1), cannot be written as the sum of two squares. If this does not happen, the number can be written as the sum of two squares.

We have now completely solved the Diophantine equation $b = x^2 + y^2$ for any positive integer **b**. We have demonstrated how the solution arises by a generalization from the prime number case, and in so doing have opened up the vast arena of number theory in all its glory.

## Stage 4

Later you might want to go back to the Sum of Squares for one final problem. Given two numbers that can be written as the sum of two squares, can their sum always be written as the sum of two squares?

For example, $10 = 3^2 + 1^2$ and $8 = 2^2 + 2^2$, and $10 + 8 = 18 = 3^2 + 3^2$. However, $20 = 4^2 = 2^2$ and $10 + 20 = 30 = 3 \times 2 \times 5$ cannot be written as the sum of two squares.

Therefore the problem is immediately solved: if two numbers can each be written as the sum of two squares, their sum may or may not be written as the sum of two squares.

On the other hand, the product of two numbers that can each be written as the sum of two squares seems to be a more promising venture. Have the children compose the following type of table (Table **13**):

Table 13. Stage 4 of sums of squares problem.

| First Number | Second Number | Product |
|---|---|---|
| $10 = 3^2 + 1^2$ | $5 = 2^2 + 1^2$ | $50 = 5^2 + 5^2$ |
| $8 = 2^2 + 2^2$ | $9 = 3^2 + 0^2$ | $72 = 6^2 + 6^2$ |
| $20 = 4^2 + 2^2$ | $4 = 2^2 + 0^2$ | $80 = 8^2 + 4^2$ |
| $90 = 9^2 + 3^2$ | $2 = 1^2 + 1^2$ | $180 = 12^2 + 6^2$ |

It seems to be the case that the product of two numbers that can each be written as the sum of two squares can also be written as the sum of two squares. Ask the children to come up with more examples on their own, and have them try to find an example that does not work out as the above table.

The problem can be settled in the affirmative; it is true that if each of two numbers can be written as the sum of two squares then their product can also be written as the sum of two squares. As a final task, we will now explore exactly how these products of numbers can be written as the sum of two squares.

## Stage 5

Repeat the original table from Stage 4 with a few more entries (Table **14**):

**Table 14. Stage 5 of sums of squares problem.**

| First Number | Second Number | Product |
|---|---|---|
| $10 = 3^2 + 1^2$ | $5 = 2^2 + 1^2$ | $50 = 5^2 + 5^2$ |
| $8 = 2^2 + 2^2$ | $9 = 3^2 + 0^2$ | $172 = 6^2 + 6^2$ |
| $20 = 4^2 + 2^2$ | $4 = 2^2 + 0^2$ | $80 = 8^2 + 4^2$ |
| $90 = 9^2 + 3^2$ | $2 = 1^2 + 1^2$ | $180 = 12^2 + 6^2$ |
| $5 = 2^2 + 1^2$ | $13 = 3^2 + 2^2$ | $64 = 8^2 + 1^2$ |
| $40 = 6^2 + 2^2$ | $5 = 2^2 + 1^2$ | $200 = 14^2 + 2^2$ |
| $25 = 5^2 + 0^2$ | $10 = 3^2 + 1^2$ | $250 = 15^2 + 5^2$ |

Ask the children if they see any way to tell what the actual squares are in the product from the squares in the first and second numbers. Hint: It has something to do with adding, subtracting, and multiplying the squares in the first two numbers in a certain way.

For example, $25 = 5^2 + 0^2$, $10 = 3^2 + 1^2$, and $25 \times 10 = 250 = 15^2 + 5^2$. How do we get 15 and 5 from 5 and 0 and 3 and 1? Perhaps someone will realize that $15 = 5 \times 3$ and $5 = 5 \times 1$.

Then look at another example: $90 = 9^2 + 3^2$; $2 = 1^2 + 1^2$; and $90 \times 2 = 180 = 12^2 + 6^2$. Someone may observe that $12 = 9 \times 1 + 3 \times 1$ and $6 = 9 \times 1 - 3 \times 1$.

Ask the children if they can see a pattern that always works. The pattern can be described more clearly using a little algebra—although algebra is not necessary in order to convey the pattern. The algebraic formulation is as follows: if $a^2 + b^2$ is the first number and $c^2 + d^2$ is the second number, the product can always be written as $(ac + bd)^2 + (ad - bc)^2$.

To describe this algebraic formula, you might say: Take the product of the first square (meaning the actual number—not the number squared) in the first number and the first square in the second number, and add this to the product of the second square in the first number and the second square in the second number; this will be the first square in the product. Then take the product of the first square in the first number and the second square in the second number, and subtract the product of the second square in the first number and the first square in the second number; this will be the second square in the product.

This non-algebraic description may be enough motivation for teaching your children a little algebra ahead of time! The formula will work whether ad − bc is positive or negative, but you might want to arrange your numbers so that ad − bc will turn out to be positive. For example, 20 x 4 = 80, 20 = $4^2 + 2^2$ and 4 = $2^2 + 0^2$, so ac + bd = 4 x 2 + 2 x 0 = 8 and ad − bc = 4 x 0 − 2 x 2 = −4.

Although it is certainly true that $8^2 + (−4)^2 = 80$, the negative number could be avoided by writing 4 = $0^2 + 2^2$; then ac + bd = 4 x 0 + 2 x 2 = 4 and ad − bc = 4 x 2 − 2 x 0 = 8.

As a final suggestion for algebra students who know how to multiply binomials, it can be mathematically proven that this formula always works—in the following way:

$$(a^2 + b^2) \times (c^2 + d^2) = a^2c^2 + a^2d^2 + b^2c^2 + b^2d^2; (ac + bd)^2 + (ad − bc)^2 = a^2c^2 + b^2d^2 + 2acbd + a^2d^2 + b^2c^2 − 2adbc = a^2c^2 + a^2d^2 + b^2c^2 + b^2d^2; \text{therefore } (a^2 + b^2) \times (c^2 + d^2) = (ac + bd)^2 + (ad − bc)^2.$$

## 10. AMICABLE NUMBERS (G)

A longtime favorite topic of both amateur and professional mathematicians is the discovery of pairs of numbers that are linked together in a certain way, known as "amicable numbers". By definition, amicable numbers are two numbers such that the sum of all the proper divisors of the first number equals the second number and the sum of all the proper divisors of the second number equals the first number.

It is not an easy problem to find a pair of amicable numbers, as the first pair of numbers with this property are both in the 200s. Like perfect numbers, prime numbers, abundant numbers, and semi-perfect numbers, finding this pair of amicable numbers is a playful way of having your students work on their long-division skills.

Suppose a child was testing to see if the number 250 was one of these amicable numbers. He/she must first come up with all the proper divisors of 250; this can be done by dividing all numbers into 250 through 15—since $16^2 = 256$. The proper divisors are 1, 2, 5, 10, 125, 50 and 25, and the sum of these numbers is 218; he/she must then find all the proper divisors of 218, testing all numbers up to 14—since $15^2 = 225$. The proper divisors of 218.

To describe this algebraic formula, you might say: Take the product of the first square (meaning the actua are 1, 2 and 109, adding up to 112. Therefore 250 and 218 are not a pair of amicable numbers. After enough trial and error—and excellent practice—somebody will find the first pair of amicable numbers to be

220 and 284.

## 11. POWERFUL NUMBERS (G, J)

Powerful numbers are numbers where all the prime numbers in their factorizations have exponents of 2 or more.

For example, 72 is a powerful number because $72 = 2^2 \times 3^2$. Any perfect square is a powerful number: $25 = 5^2$, $16 = 4^2$; 400 is a powerful number because $400 = 5^2 \times 4^2$ as is $64 = 2^6$. However, 180 is not a powerful number, because $180 = 2^2 \times 3^2 \times 5$.

At first glance, finding powerful numbers seems to involve nothing more than determining the unique prime factorization of a number.

However, try finding two consecutive powerful numbers. Hint: There are two consecutive single-digit powerful numbers. It should not take very long for someone to come up with $8 = 2^3$ and $9 = 3^2$. However, the problem of finding two other consecutive powerful numbers is not so immediate.

Tell the students the next smallest pair of consecutive powerful numbers is in the 200s. After some valuable trial and error prime factorization work, someone should determine that the answer to the problem is $288 = 2^5 \times 3^2$ and $289 = 17^2$.

## 12. KAPREKAR NUMBERS (D)

Children can practice two-digit multiplication, in particular finding squares of numbers, using Kaprekar numbers. The best way to understand these numbers is to give an example: $45^2 = 2025$ and $20 + 25 = 45$; therefore, 45 is a Kaprekar number.

Take the square of a two-digit number and add the two pairs of numbers in your answer. If this sum equals the number you began with, the number is called a Kaprekar number.

If a number has only three digits in its square, such as $31^2 = 961$, write the square as 0961 and pairs of digits as 9 and 61. There are three two-digit Kaprekar numbers, and, after a little work, all your students should be able to come up with the other two: $55^2 = 3025$ and $99^2 = 9801$.

## 13. PASCAL'S TRIANGLE AND TRIANGULAR NUMBERS (A, C, D, F, G, J)

Pascal's Triangle is a useful array of numbers.

## Stage 1

Draw the following array of numbers; place the number 1 at the end of each row; take the sum of consecutive numbers in a row and write this sum midway below the two numbers (Table **15**).

Table 15. Stage 1a of pascal's triangle and triangular numbers problem.

```
                    1
                   1 1
                  1 2 1
                 1 3 3 1
                1 4 6 4 1
              1 5 10 10 5 1
            1 6 15 20 15 6 1
          1 7 21 35 35 21 7 1
        1 8 28 56 70 56 28 8 1
    1 9 36 84 126 126 84 36 9 1
```

As evident, this array of numbers forms a triangle that increases in size with each subsequent row. Note that the numbers in each row follow the same sequence from the left as from the right.

The diagonals on the outside are all ones; the second diagonals are the ordinary counting numbers: 1, 2, 3, ...; and the third diagonals are the sequence of numbers: 1, 3, 6, 10, 15, ...

Ask the children what the pattern is by having them come up with the next number in the sequence.

Each number is obtained by adding one more than the difference of the previous two numbers—to the last number. These numbers are known as "triangular numbers".

Note that these numbers can indeed be described geometrically as triangles, as can be seen in Table **16**:

Table 16. Stage 1b of pascal's triangle and triangular numbers problem.

| * | * | * | * | |
|---|---|---|---|---|
| | ** | ** | ** | |
| | | *** | *** | |
| | | | **** | |
| 1 | 3 | 6 | 10 | etc. |

Next, have the children add the numbers in each row. They will see that the sum seems to double.

You might play the game of reminding them about what happened with the doubling phenomena back in the problem about subsets and circles.

Ask them to guess if this doubling pattern always continues, having them add 10 rows or 12 rows—or as many as you would like them to add.

An even more unusual feature of Pascal's triangle has to do with division.

Ask the children to see if the row number itself (except for row 0) always divides evenly (without a remainder), into each of the numbers in the row, other than 1.

For example, this works for rows 2 and 3 (there is nothing for row 1 to divide), but does not work for row 4, since 4 does not divide evenly into 6. Row 5 works, row 6 doesn't, row 7 works, row 8 doesn't, *etc*. Make a table that looks like Table **17**:

**Table 17. Stage 1c of pascal's triangle and triangular numbers problem.**

| Yes | No |
|-----|-----|
| 2 | 4 |
| 3 | 6 |
| 5 | 8 |
| 7 | |

Ask the children if they can discover a pattern to determine which rows work and which rows do not. This is excellent division practice, and important discoveries shall soon be made—as row 9 turns out to be a No, thereby foiling the potential hypothesis that all odd numbers are Yes's.

Continuing the table a little further, it will be found that 10 is a No, 11 a Yes, 12 a No, and 13 a Yes.

Is it clear what the correct solution is? It is our old friend—prime numbers; all prime numbers are Yes's and all composite (non-prime) numbers are No's.

## Stage 2

Now that triangular numbers have been introduced, you might try exploring the relationship between triangular numbers and perfect squares. Have the children list the first 10 triangular numbers and the first 10 perfect squares, making a table that looks like Table **18**:

**Table 18. Stage 2a of pascal's triangle and triangular numbers problem.**

| Triangular Numbers | Perfect Squares |
|:---:|:---:|
| 1 | 1 |
| 3 | 4 |
| 6 | 9 |
| 10 | 16 |
| 15 | 25 |
| 21 | 36 |
| 28 | 49 |
| 36 | 64 |
| 45 | 81 |
| 55 | 100 |

Ask them what is the relationship between triangular numbers and perfect squares. Hint: it has something to do with addition.

The sum of any two triangular numbers equals a perfect square: $3 + 6 = 9 = 3^2$; $6 + 10 = 4^2$; $10 + 15 = 25 = 5^2$, *etc.* You might have them continue the table until $20^2 = 400$ to see if this pattern continues. Algebraically the relationship can be stated as follows: If $T_1$, $T_2$, $T_3$, ... $T_n$ are the first **n** triangular numbers, then $T_i + T_{i+1} = (i + 1^2)$ where $1 \le i \le n - 1$.

Another interesting relationship exists between squares of triangular numbers and cubes of ordinary numbers. Make a table that looks like Table **19**:

**Table 19. Stage 2b of pascal's triangle and triangular numbers problem.**

| Number | Triangular Number | Square of Triangular Number | Number Cube |
|:---:|:---:|:---:|:---:|
| 1 | 1 1 | 1 | |
| 2 | 3 | 9 | 8 |
| 3 | 6 | 36 | 27 |
| 4 | 10 | 100 | 64 |
| 5 | 15 | 225 | 128 |

*(Table 19) contd.....*

| 6 | 21 | 441 | 216 |
|---|---|---|---|

Ask the children if they can find the relationship that exists between triangular squares and number cubes. You might have them continue the table through the number 10.

The relationship is that the sum of any number of consecutive cubes (starting from the first cube) always equals the square of a triangular number. Stated algebraically, $1^3 + 2^3 + 3^3 \dots + n^3 = (t_n)^2$. For example, $1 + 3 + 27 = 36 = 6^2$, $1 + 8 + 27 + 36 + 64 + 125 + 216 = 441 = 21^2$, *etc.* This is good practice in multiplication, addition, and excellent insight into some of the mysteries that are inherent in our number system.

**Stage 3**

This next stage of Pascal's Triangle is primarily for a student learning algebra. There is a neat algebraic formula for obtaining the nth triangular number.

Notice that the first triangular number, 1, can be written (1 x 2)/2. The second triangular number, 3, can be written as (2 x 3)/2 The third, 6, can be written (3 x 4)/2, and the tenth triangular number, 55, can be written as (10 x 11)/2.

Ask the children to write down the general algebraic formula for obtaining the nth triangular number. It is (n x (n + 1)) ÷ 2.

A final problem that all algebra students ought to become familiar with soon after learning how to multiply two binomials and trinomials is the following:

Start with $(x + 1)^2 = x^2 + 2 + 1$. Then continue increasing the exponents and painstakingly work out the multiplication for a while:

$$(x + 1)^3 = (x + 1)(x + 1)^2 = x^3 + 3x^2 + 3x + 1.$$
$$(x + 1)^4 = (x + 1)^2(x + 1)^2 = x^4 + 4x^3 + 6x^2 - 4x + 1,$$
$$(x + 1)^5 = (x + 1)(x + 1)^4 = x^5 + 5x^4 + 10x^3 + 10x^2 + 5x + 1.$$

Then look at the first six rows of Pascal's Triangle:

Your students will discover that the numbers in each row of Pascal's Triangle are exactly the coefficients of the corresponding binomial expansion.

Have them verify that this relationship indeed continues for rows 6, 7, and 8.

**14. PRIME FINDING FUNCTIONS (G, J)**

This problem, though basically a problem in elementary algebra for middle school students, also serves as a good preview of algebra for bright elementary school students. Ask your students to pick a counting number from zero to 10, square it, add the number, and then add 17 (Table **20**):

**Table 20. Stage 3 of pascal's triangle and triangular numbers problem.**

<pre>
            1
           1 1
          1 2 1
         1 3 3 1
        1 4 6 4 1
      1 5 10 10 5 1
    1 6 15 20 15 6 1
</pre>

For example, if they were to pick the number 8, they would get $8^2 + 8 + 17 = 64 + 8 + 17 = 89$. Make a table that looks like Table **21**:

**Table 21. Table for prime finding functions.**

| Beginning Number | Ending Number |
|---|---|
|  | 17 |
| 1 | 19 |
| 2 | 23 |
| 3 | 29 |
| 4 | 37 |
| 5 | 47 |
| 6 | 59 |
| 7 | 72 |
| 8 | 89 |
| 9 | 107 |
| 10 | 127 |

Ask the children what all the numbers in the second column have in common and undoubtedly by now someone will realize that they are all prime numbers. (Note that the difference between two successive numbers is two more than the difference of the two previous successive numbers.) The question to ask is: Does the procedure always result in prime numbers, or does it eventually break down and form a composite number?

There is only one way to find out, and we have previously seen, testing for primes is an excellent way to practice division. The procedure yields prime numbers through 15, but fails to work for 16 as $16^2 + 16 + 17 = 289 = 17^2$; algebraically, we are working with polynomials of the form $x^2 + x + 17$; the smallest value of x that results in a composite number is x = 16.

Notice that when x is 16, the composite number is $17^2 = 289$. This is not just a coincidence; the same phenomenon results from $x^2 + x + 3$, $x^2 + x + 5$, $x^2 + x + 11$ and, surprisingly, not again until $x^2 + x + 41$.

For this last polynomial notice that when x = 40, $40^2 + 40 + 41 = 1600 + 40 + 41$ = 1681 = $41^2$. The reason for this phenomenon involves the world of pure mathematics.

## 15. EULER PHI FUNCTIONS (G, J)

### Stage 1

Two numbers are defined to be "relatively prime" if there are no numbers that divide evenly into both of them—other than 1.

For example, 9 and 10 are relatively prime, as are 8 and 15, *etc.* Given any number we can count exactly how many numbers are relatively prime to it. For 8, we would have four relatively prime numbers: 1, 3, 5, and 7. We use the symbol $\emptyset(8) = 4$ to say that phi of 8 = 4.

The number of numbers relatively prime to a given number is mathematically known as the Euler Phi function. Next, take phi of any prime number such as b. The numbers relatively prime to 5 are 1, 2, 3 and 4; and therefore $\emptyset(5) = 4$. Notice that $\emptyset(3) = 2$, $\emptyset(7) = 6$, $\emptyset(11) = 10$, *etc.*

All prime numbers have the phi value one less than the given number. Algebraically, we can state that $\emptyset(p) = p - 1$ if p is a prime number. At any rate, we know that $\emptyset(8) = 4$ and $\emptyset(5) = 4$.

If we multiply 8 x 5 we get 40. What do you think happens if we take $\emptyset(40)$? The numbers that are relatively prime to 40 are 1, 3, 7, 9, 11, 13, 17, 19, 21, 23, 27, 29, 31, 33, 37, and 39.

There are 16 of these numbers. Notice that $\emptyset(8 \times 5) = \emptyset(40) = 16$ and $\emptyset(8) \times \emptyset(5)$ = 4 x 4 = 16, so we can say that $\emptyset(8 \times 5) = \emptyset(8) \times \emptyset(5)$.

Is this just a chance phenomenon or does this type of thing always occur? Have your students offer a few more numbers for consideration and begin making a table like Table **22**:

**Table 22. Table 1 for Euler phi functions problem.**

| Yes | No |
|---|---|
| Ø(8 x 5) = Ø(8) x Ø(5) | |

Let's suppose you were given the numbers 7 and 4, and Ø(7) = 6 and Ø(4) = 3 and Ø(7 x 4) = Ø(28) which does indeed equal 18, so 7 x 4 = 28 is a yes. However, the process will break down with the numbers 4 and 6: Ø(4) = 2 and Ø(6) = 5, but Ø(4 x 6) ≠ Ø(4) x Ø(6). (The numbers relatively prime to 24 are 1, 5,7, 11, 13, 17, 19, and 23 so Ø(24) = 8 ≠ 5 x 2). Our table now looks like Table **23**:

**Table 23. Table 2 for Euler phi functions problem.**

| Yes | No |
|---|---|
| 8 x 5 = 40 | 4 x 6 = 24 |
| 7 x 4 = 28 | |

With a few more entries the table will grow in size and useful information and perhaps might look something like Table **24**:

**Table 24. Table 3 for Euler phi functions problem.**

| Yes | No |
|---|---|
| 8 x 5 = 40 | 4 x 6 = 24 |
| 7 x 4 = 28 | 9 x 3 = 27 |
| 5 x 6 = 30 | 10 x 4 = 40 |

This is an excellent problem in discovery mathematics for your students. From the table, it appears that two even numbers are a No, such as 4 x 6 and 10 x 4, and a number with a multiple of itself is a No, such as 3 and 9.

However, it can also be checked that 9 x 6 = 54 is a No, and it turns out that the simple elegant solution is as follows: *Whenever two numbers are relatively prime the phenomenon takes place; when two numbers are not relatively prime the phenomenon does not take place.*

It turns out that this relatively prime requirement is fairly common in number theory. Algebraically we can state that Ø(a x b) = Ø(a) x Ø(b) if and only if a and b are relatively prime.

We call a function that has this property—such as Ø —a multiplicative function.

## Stage 2

A further development of the Euler Phi function can be explored in the following way. Pick a number and find all the divisors of the number. For example, if we pick 12, the divisors of 12 are 1, 2, 3, 4, 6, and 12. For each of these divisors, apply the Euler Phi function (*i.e.*, determine how many numbers are relatively prime to each one of the divisors.) Thus $\varnothing(1) = 1$, $\varnothing(2) = 1$, $\varnothing(3) = 2$, $\varnothing(4) = 2$, $\varnothing(6) = 2$, and $\varnothing(12) = 4$.

Adding up all these Phi numbers yields $1 + 1 + 2 + 2 + 2 + 4 = 12$, which is our original number. Quite amazingly, this procedure will always work. Have your students try it out for other numbers and verify that it is true.

Taking one more example, the divisors of 16 are 1, 2, 4, 8, and 16, $\varnothing(1) = 1$, $\varnothing(2) = 1$, $\varnothing(4) = 2$, $\varnothing(8) = 4$, and $\varnothing(16) = 8$; $1 + 1 + 2 + 4 + 8 = 16$.

The algebraic way of stating this property makes use of the summation symbol $\sum$. The formula is $\sum \varnothing(d_i) = n$, which means that the sum of the Euler Phi functions of all the divisors of a number equals the number.

## 16. DIVISOR FUNCTION (G, J)

To see another example of a multiplicative function (without telling your students this beforehand), define the "divisor function" of a number as the number of divisors of a given number, including the given number itself. For example, we know the divisors of 8 are 1, 2, 4, and 8 and we say $d(8) = 4$. Similarly, the divisors of 9 are 1, 3, and 9 so $d(9) = 3$.

Since any prime number can have only two divisors, the number itself and 1, $d(p) = 2$ for any prime number p. Notice that $d(72) = 12$ since 72 has 12 divisors: 1, 2, 3, 4, 6, 8, 9, 12, 18, 24, 36, and 72.

Notice also that since $d(8) = 4$ and $d(9) = 3$, we have $d(8) \times d(9) = 4 \times 3 = 12$ and $d(8 \times 9) = d(8) \times d(9)$.

Spread the Euler Phi function and divisor function discussions with a few different problems in between. Start making a table from scratch for the divisor function in a similar way to what was done for the Euler Phi function and have your students discover the criteria for themselves.

## 17. GOLDBACH CONJECTURE (F, G)

This problem is one of the most innocently annoying problems in the history of mathematics. It is simple to formulate and understand but absurdly difficult to solve.

The Goldbach Conjecture states that every even number other than 2 can be written as the sum of two prime numbers. For example, $8 = 5 + 3$, $14 = 11 + 3$, $38 = 31 + 7$, $102 = 97 + 5$, *etc.*

It certainly appears as if the Goldbach Conjecture is true, (*i.e.*, that every even number other than 2 can be written as the sum of two prime numbers.)

Have your students verify that the Goldbach Conjecture works for all even numbers up to 200—or 300. The mystery is that no one has ever been able to prove that it works for all even numbers and no one has been able to come up with an even number for which the Goldbach Conjecture does not work.

Here is a wonderful example of an unsolved mathematical problem—for over 250 years thus far—that your students can well appreciate—and even work on!

## 18. SUMS OF FOUR SQUARES (D)

Recall that a number can be written as the sum of two squares if and only if there is no prime of the form $4k + 3$ to an odd exponent in its prime factorization. The next question we might think of asking is: Can a number always be written as the sum of three squares? For example, 11 cannot be written as the sum of two squares, but $11 = 3^2 + 1^2 + 1^2$ and therefore can be written as the sum of three squares.

Similarly, 30 cannot be written as the sum of two squares ($30 = 3 \times 2 \times 5$) but $30 = 5^2 + 2^2 + 1^2$ so it can be written as the sum of three squares. However, it should not take too long before one of your students determines that not all numbers can be written as the sum of three squares either. 7 needs at least four squares: $2^2 + 1^2 + 1^2 + 1^2$; so does 15 ($15 = 3^2 + 2^2 + 1^2 + 1^2$); and 23 ($23 = 3^2 + 3^2 + 2^2 + 1^2$).

The next obvious question is: Can all numbers be written as the sum of, at most, four squares? A good exercise is to have your students make a list of all two-digit numbers that require at least four squares to equal the number.

The list of numbers through 60 that have this property are 7, 15, 23, 28, 31, 39, 47, 55, and 60.

It has been proven that every number can indeed be written as the sum of, at most, four squares, no matter how large the number is.

## 19. FIBONACCI NUMBERS (A, J)

### Stage 1

I like to begin this sequence of numbers by discussing the family tree of a male honeybee. A male honeybee comes from only one parent—a female—while a female honeybee comes from two parents—both a male and a female. You can therefore construct a family tree for a male honeybee that looks like the following, with ♀ as a female bee and ♂ as a male bee (Fig. **5**).

The numbers on the right are the number of bees in each generation. Ask the children if they see a pattern for how many bees are in each generation.

It is likely that someone will think that there will be 12 bees in the next generation, as 1, 2, 3, 5, 8 does resemble the pattern of triangular numbers where the difference of two successive numbers increased by one; thus 8 + 4 = 12. However, there are 13 bees in the next generation, and notice that when two successive numbers in the sequence are added, you get the next number.

You can decide how many generations of bees should be drawn to verify this hypothesis—21 bees, 34 bees, or 55 bees—depending upon how much you and your students enjoy drawing bees. We have now witnessed the mathematical birth of the Fibonacci numbers, found in a variety of settings in nature: bee generations, rabbit births, flower arrangements, *etc*. (Fig. **5**).

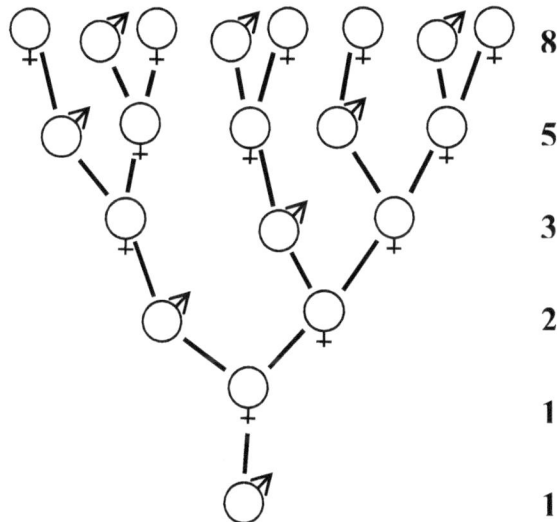

**Fig. (5).** Stage 1 of Fibonacci Numbers Problem.

## Stage 2

There are a number of algebraic relationships that exist within the realm of the Fibonacci numbers. Make a table of the first 10 Fibonacci numbers alongside their corresponding sums, as follows (Table **25**):

Table 25. Stage 2 of Fibonacci numbers problem.

| Fibonacci Numbers | Corresponding Sums |
|:---:|:---:|
| 1 | 1 |
| 1 | 2 |
| 2 | 4 |
| 3 | 7 |
| 5 | 12 |
| 8 | 20 |
| 13 | 33 |
| 21 | 54 |
| 34 | 88 |
| 55 | 143 |

Ask your students to determine the relationship that exists between these two columns. One thing that might be observed is that the column on the right happens to be one less than each Fibonacci number, starting with one less than the third Fibonacci number, 2.

In other words each Fibonacci number in the column on the left (beginning with the third row) happens to be one more than the corresponding number that is two rows above it in the column on the right.

Also, the sum of the first three Fibonacci numbers, $1 + 1 + 2 = 4$, is one less than the fifth Fibonacci number; continuing in this way the sum of the first eight Fibonacci numbers, $1 + 1 + 2 + 3 + 5 + 8 + 13 + 21 = 54$, is one less than the tenth Fibonacci number, 55.

The general algebraic relationship is $F_1 + F_2 + ... + F_n = F_{n+2} - 1$. There are a number of other interesting formulas that can be found with the Fibonacci numbers, some of which are described in problem #44, *More on Fibonacci Numbers*.

There is also a relationship between the Fibonacci sequence to another sequence of numbers called the Lucas sequence.

The Lucas sequence is 1, 3, 4, 7, 11, 18, 29, 47, ... These numbers are obtained similarly to the Fibonacci sequence: The sum of two consecutive numbers in the Lucas sequence yields the next number.

Make a table listing the first 10 Fibonacci numbers and the first 10 Lucas numbers. Call them F and L numbers.

Ask the children to determine the relationship between these two sequences.

The solution is that the sum of alternate Fibonacci numbers always equals the intermediate Lucas number. For example, the third Fibonacci number is 2, the fifth Fibonacci number is 5, and 2 + 5 = 7, which is the fourth Lucas number. For another example, the eighth Fibonacci number is 21, the tenth Fibonacci number is 55, and 21 + 55 = 76, which is the ninth Lucas number. Algebraically this can be succinctly stated as $L_n = F_{n-1} + F_{n+1}$ .

## Stage 3

A favorite problem for algebra students looks at the relationship of the Fibonacci sequence to the Golden Ratio. This problem does require a fairly good background in high school algebra—including a knowledge of algebraic fractions, simplifying radicals, and the Quadratic Formula.

To begin, define the Golden Ratio as the ratio of the length to the width of the rectangle with the following property (Fig. **6**):

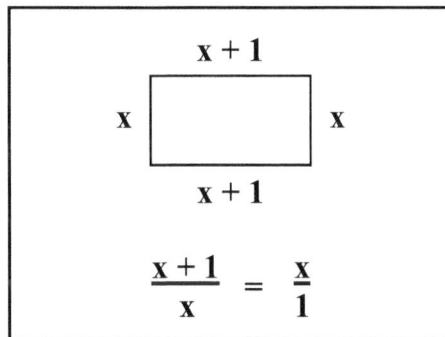

**Fig. (6).** Stage 3 of Fibonacci Numbers Problem.

Cross multiplying, we set up the quadratic equation: $x^2 = x + 1$ or $x^2 - x - 1 = 0$, which can easily be solved using the Quadratic Formula. The solution is found to be $x = (1 \pm \sqrt{5}) \div 2$.

We will just use the approximate positive value x $\cong$ 1.68. (The actual number is irrational and has an infinite decimal expansion).

Writing out the first 10 Fibonacci numbers, 1, 1, 2, 3, 5, 8, 13, 21, 34, 55, take ratios of successive numbers and you will get 1/1 = 1, 2/1 = 2, 3/2 = 1.5, 5/3 = 1.666 ..., 8/5 = 1.6, 13/8 = 1.625, *etc.*

Notice that these ratios seem to be approaching the Golden Ratio. Indeed this is exactly what happens, and we say that the limit of the ratios of successive numbers of the Fibonacci sequence is the Golden Ratio (a preview of Calculus). But the real fun starts when we try to go the other way.

Beginning with the Golden Ratio x = $(1 \pm \sqrt{5}) \div 2$, take the square of this number: $x^2 = (3 + \sqrt{5}) \div 2$. (Notice that this equality actually says $x^2 = x + 1$, which is how we began).

Now take $x^3 = xx^2$. If we continue this procedure for higher powers of x, we will obtain: $x^4 = 3x + 2$, $x^5 = 5x + 3$, $x^6 = 8x + 5$, *etc.* It should be obvious that both the coefficient of the x term and the constant term itself are forming the Fibonacci sequence, the coefficient of the x term being one Fibonacci term higher than the constant term. Verifying this for $x^4$, $x^5$, $x^6$, *etc.*, is an excellent exercise for algebra students studying simplification by radicals and algebraic fractions. At any rate, we have now seen the remarkable relationship that exists between the Fibonacci sequence and the Golden Ratio.

## 20. NUMBER OF DIVISORS (G, J)

There is a wonderful way of determining how many divisors (*i.e.*, how many numbers divide evenly into a given number). Take a number, such as 30; it has the following divisors: 1, 2, 3, 5, 6, 10, 15, and 30 (or 8 divisors); 32 has 1, 2, 4, 8, 16, and 32 (6 divisors); 100 has 1, 2, 4, 5, 10, 20, 25, 50, and 100 (9 divisors). Now write down the prime factorization of these numbers, as Table **26**:

Table 26. Table 1 for number of divisors problem.

| Number | Prime Factorization | Number of Divisors |
|--------|---------------------|--------------------|
| 30 | 3 x 2 x 5 | 8 |
| 32 | $2^5$ | 6 |
| 100 | $2^2$ x $5^2$ | 9 |

The mystery is that the number of divisors has something to do with the prime factorization. Hint: It has something to do with the exponents of the prime numbers, not the prime numbers themselves. We'll think of 3 x 2 x 5 as $3^1$ x $2^1$ x

$5^1$; thus these exponents are all ones. A few examples might be helpful (Table **27**):

**Table 27. Table 2 for number of divisors problem.**

| Number | Prime Factorization | Number of Divisors |
|--------|---------------------|--------------------|
| 30 | $3^1$ x $2^1$ x $5^1$ | 8 |
| 32 | $2^5$ | 6 |
| 100 | $2^2$ x $5^2$ | 9 |
| 50 | $5^2$ x 2 | 6 |
| 11 | $11^1$ | 2 |
| 80 | $2^4$ x $5^1$ | 10 |

This isn't an easy problem. Tell your students to look at what happens for prime numbers. Every prime number has only two divisors, itself and 1, and each has an exponent of 1.

Next, suggest that they look at powers of primes. For example, $32 = 2^5$ has 6 divisors—with an exponent of 5; $25 = 5^2$ has 3 divisors—with an exponent of 2; $27 = 3^3$ has 4 divisors—with an exponent of 3.

The number of divisors for primes and powers of primes is always 1 more than the prime number's exponent. The answer is to increase each exponent by 1 and multiply the new exponents together in order to get the number of divisors for any number.

Some examples are: $30 = 3^1$ x $2^1$ x $5^1$ and has 2 x 2 x 2 = 8 divisors; $100 = 2^2$ x $5^2$ and has 3 x 3 = 9 divisors; $80 = 2^4$ x $5^1$ and has 5 x 2 = 10 divisors.

## 21. EULER'S FORMULA (G, J)

Choose a prime number and take any other number that is not a multiple of this number. For example, 7 and 2, or 5 and 9, or 3 and 4, *etc.*

Now take 1 less than your original prime number, and use these new numbers as exponents for your original second number, as in Table **28**:

**Table 28. Table 1 for Euler's formula problem.**

| 1 Less than Prime | Second Number | Number to Exponent |
|-------------------|---------------|--------------------|
| 7 – 1 = 6 | 2 | $2^6 = 64$ |
| 5 – 1 = 4 | 4 | $4^4 = 256$ |
| 3 – 1 = 2 | 4 | $4^2 = 16$ |

Subtract 1 from each of the numbers in the third column, yielding, respectively, 63, 255 and 15. The original prime numbers divide evenly into each of these three numbers respectively, (*i.e.*, 7 divides evenly into 63, 5 into 255, and 3 into 15). This phenomenon can be described more completely as in Table **29**—using more examples.

Table 29. Table 2 for Euler's formula problem.

| Prime Number | Second Number | 1 Less Than Number to Exponent | Division |
|---|---|---|---|
| 7 | 2 | $2^6 - 1 = 63$ | $63/7 = 9$ |
| 5 | 4 | $4^4 - 1 = 255$ | $255/5 = 51$ |
| 3 | 4 | $4^2 - 1 = 15$ | $15/3 = 5$ |
| 11 | 2 | $2^{10} - 1 = 1023$ | $1023/11 = 9$ |
| 2 | 13 | $13^1 - 1 = 12$ | $12/2 = 6$ |

Have your students try this procedure for various pairs of numbers, verifying that it always works. This is excellent practice in using exponents and division, and in following a step-by-step mathematical procedure.

Algebraically, the problem can be stated in the following way: If the greatest common divisor of a prime p and another number, denoted by "a," is 1, written as (a, p) = 1, then $a^{p-1} - 1$ is divisible by a. This is known as "Euler's Formula".

## 22. NUMBER REVERSALS (D)

Ask your students to square 12, getting 144. Now ask them to reverse the digits of 12, getting 21, and square 21, getting 441. Notice that 441 is the reverse of 144. Have them find another two-digit number that has this same square-reversal property. If they go about it systematically, they should hit upon 13. Note that $13^2 = 169$ and $31^2 = 961$. There is no other two-digit number with this square reversal property.

## 23. ONE-TWO-THREE NUMBERS (D)

Take any three-digit number, write down the digit on the left, the square of the digit in the middle, and the cube of the digit on the right. Add these numbers, and if you obtain your original number, call it a one-two-three number.

For example, test 164. You obtain $1 + 6^2 + 4^3 = 1 + 36 + 64 = 101$; therefore, 164 is not a one-two-three number.

Have your students try to find a one-two-three number: You might tell them there is one less than 200. The correct answer is 135: $1 + 3^2 + 5^3 = 1 + 9 + 125 = 135$.

## 24. ANOMALOUS FRACTIONS (H, J)

Now we come to a humorous change of pace that is perfect for students who have recently become adept at reducing fractions to lowest terms.

### Stage 1

Have the children reduce 16/64 to lowest terms: Let them do it step-by-step in any way they like. One possibility is 16/64 = 8/32 = 4/16 = 2/8 = 1/4. Now point out to them (and to yourself) that they could have reduced 16/64 to lowest terms much more easily by simply cancelling the 6's: 16/64 = 1/4!

At this point, ask them to reduce 19/95 to lowest terms. Yes, canceling the 9s yields 19/95 = 1/5, which is correct.

Give them 49/98 and have them reduce their result after canceling the 9s, which will yield 49/98 = 4/8 = 1/2, which is also correct.

Now the fun starts. Ask them if they think this method always works. After trying a number of other fractions that are proper (numerator less than denominator) and two digit with the ones digit in the numerator the same as the tens digit in the denominator, (*e.g.*, 15/58, 28/84, 36/62, *etc.*), their enthusiasm for this unusual ad hoc method of reducing fractions to lowest terms will most certainly wane.

Ask them how many two-digit proper fractions they believe this method works for, and someone should answer "4," which is correct. The students' job (and your own) is to find the fourth fraction that has this property.

These fractions are called "anomalous". Please forgive me for not telling you the answer to this problem. After all, you might as well experience a bit of the adventure yourself!

### Stage 2

The above problem can be formulated algebraically in the following way. Consider 16/64 as:

$$\frac{10 \text{ x } 1 + 6}{10 \text{ x } 6 + 4} = 1/4$$

In general, uv/vw means

$$\frac{10u + v}{10v + w}$$

Cross multiplying yields w(10u + v) = u(10v + w) or 10uw + wv = 10uv + uw. For the fractions to be proper, either u < v or u = v and v < w. Now check the three anomalous fractions: 16/64, 19/95 and 49/98.

For 16/64, u = 1, v = 6, w = 4, and 4 x (10 x 1 + 6) = 1 x (10 x 6 + 4) and 4 x 16 = 64, 1 x 64 = 64.

For 19/95, u = 1, v = 9, w = 5 and 5 x (10 x 1 + 9) = 1 x (10 x 9 + 5) and 5 x 19 = 95, 1 x 95 = 95.

For 49/98, u = 4, v = 9, w = 8 and 8 x (10 x 1 + 9) = 4 x (10 x 9 + 8) and 8 x 49 =392, 4 x 98 = 392.

The fourth and final anomalous fraction will also satisfy this algebraic equation. There is a generalization of anomalous fractions to base systems other than 10. The interested reader can find more information about this in the book *Solving Math Problems in Basic* listed in the Bibliography.

## 25. AUTOMORPHISMS (D)

Multiplication in terms of the square of a number may be an "automorphism". 25 is an automorphism because $25^2 = 625$, and the digits 25 are the last two digits of the answer. Ask your students to find another two-digit number that is an automorphism. They should come up with 76, as $76^2 = 5776$.

## 26. SUM OF DIGITS NUMBERS (A)

Here is a problem that requires no skill other than two-digit addition and persistence. Take a three-digit number, 156 for example. Form all of the possible two-digit numbers. out of this number, without using the same number twice.

In the case of 156, the numbers would be 15, 16, 51, 56, 65, and 61. Add these numbers. You will find that the sum is greater than the original number, 156.

Now ask your students to find a three-digit number such that the sum of all the two-digit numbers that can be formed from it add up to the number. Hint: The number is between 100 and 200 (or 100 and 150 if you prefer).

The correct answer is 132. The two-digit numbers are 13, 12, 31, 32, 21, and 23.

## 27. PRIMES AND MULTIPLES OF 6 (G)

Although we really do not know any pattern to describe prime numbers, we do know a certain formula that holds true for nearly all prime numbers. Take any prime number—other than 2 and 3—such as 17, and divide it by 6. Notice that the remainder is 5.

Take another prime number, such as 31, and divide it by 6. Notice the remainder now is 1.

Have your students continue this procedure for all primes less than 100 (or 150, or 200). For example, 79 is a prime and 79 divided by 6 leaves a remainder of 1. Another prime, 101, divided by 6 leaves 5.

It is true that every prime number—other than 2 and 3—can be characterized as leaving a remainder of 1 or 5 when divided by 6. This fact is not quite as surprising as it might seem at first glance, since every prime number—other than 2—is odd, and when an odd number is divided by an even number, the remainder is always odd.

The possible remainders, therefore, when dividing a prime number—other than 2—by 6, are 1, 3, or 5. However, the converse of this statement is not necessarily true, (*i.e.*, just because a number divided by 6 leaves a remainder of 1 or 5 does not mean that the number is prime).

For example, 25 divided by 6 leaves a remainder of 1, and 25 is certainly not prime; 35 divided by 6 leaves a remainder of 5 and likewise 35 is not prime.

At any rate, it will be a good exercise in division for your students to test a list of numbers to see if they are prime, using this remainder-when-divided-by-6 criterion as a helpful device.

Also, point out to your students that if a number is divided by 6 and leaves a remainder of 3, then it must be a multiple of 3, and, therefore, certainly cannot be a prime number. This is an example of an informal mathematical proof.

Notice also that if a number is divided by 6 and leaves a remainder of 5, that number is one less than a multiple of 6. For example, $35 = 6 \times 6 - 1$; $59 = 6 \times 10 - 1$, *etc*.

Algebraically, we say that all prime numbers with the exception of 2 and 3 are of the form $6 \times n - 1$ or $6 \times n + 1$.

## 28. CLOCK ARITHMETIC (G, J)

Now we come to a wonderful problem that utilizes the mathematical disciplines of group theory and number theory with a concrete example taken from clocks.

**Stage 1**

Draw an ordinary 12-hour clock (Fig. **7**):

**Fig. (7).** Stage 1 of Clock Arithmetic Problem.

Ask the children to do some arithmetic on the clock, such as $2 + 6 = 8$, $3 + 9 = 12$, but also $7 + 6 = 1$, $8 + 7 = 3$, $9 + 9 = 6$, *etc.* This is what is described mathematically as a modulo 12 system.

Subtracting multiples of 12 always yields the same answer. For example, $8 \times 7 = 56 - 12 \times 4 = 8$; $6 \times 8 = 48 - 12 \times 3 = 12$. Notice that 12 observes the same function as zero, since $12 + n = n$ and $12 \times n = 12$ for any number n.

For example, $12 + 5 = 17 - 12 \times 1 = 5$ and $12 \times 5 = 60 - 12 \times 4 = 12$. The problem of interest is, given any number on the clock, can you find a number which, when multiplied by that number, will yield 1.

For example, $5 \times 5 = 25 - 12 \times 2 = 1$; $7 \times 7 = 49 - 12 \times 4 = 1$; but there is no such way of multiplying 2 by any number on a 12-hour clock to obtain 1. Nor can this be done for 3, 4, 6, 8, 9, 10 or 12.

This can easily be checked. For 2, the only possibilities we can get are the even numbers on the clock: 2, 4, 6, 8, 10, and 12, thereby never getting I.

We say, therefore, that 5 and 7 have "multiplicative inverses"—meaning that there are numbers which, when multiplied by these numbers, yield 1.

We say that 2, 3, 4, 6, 8, 9, 10, and 12 do not have multiplicative inverses. A mathematical way of describing the situation is to write 5 x 5 ≡ 1 mod 12, 7 x 7 ≡ 1 mod 12 and 2 x n ≡ 1 mod 12 for any number n.

## Stage 2

Now we generalize our clock arithmetic to a 7-hour clock—as in the following diagram (Fig. **8**):

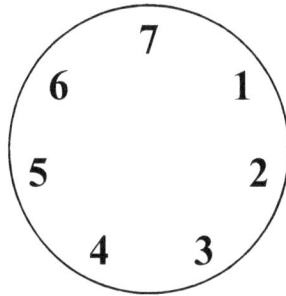

**Fig. (8).** Stage 2 of Clock Arithmetic Problem.

Addition and multiplication factors work out in a similar fashion to a 12-hour clock: 2 + 6 = 1, 5 + 8 = 6, 5 x 6 = 30 − 7 x 4 = 2, and this can again be described mathematically as 2 + 6 = 8 ≡ 1 mod 7, 5 + 8 = 13 ≡ 6 mod 7, and 5 x 6 = 30 ≡ 2 mod 7.

Notice that there are only really 6 numbers to check for multiplicative inverses since 7 times any number always gets back to 7. Test each number from 1 to 6. It is not hard to get the following results: 1 x 1 = 1 ≡ 1 mod 7, 2 x 4 = 8 ≡ 1 mod 7, 3 x 5 = 15 ≡ 1 mod 7, 4x2 = 8 ≡ 1 mod 7, 5 x 3 = 15 ≡ 1 mod 7, and 6 x 6 = 36 ≡ 1 mod 7.

Notice that every number—other than the clock number 7—has a multiplicative inverse on a 7-hour clock. Thus we see that a 7-hour clock behaves differently from a 12-hour clock.

Have the children begin testing all different clocks for multiplicative inverses. Start with a 2-hour clock and ask them to go all the way up to an 11-hour clock. If all numbers on the clock—other than the clock number itself—have multiplicative inverses, label the clock as a Yes; otherwise label it as a No. They should end up with Table **30**:

**Table 30. Stage 2 of clock arithmetic problem.**

| Yes | No |
|:---:|:---:|
| 2 | 4 |
| 3 | 6 |
| 5 | 8 |
| 7 | 10 |
| 11 | 12 |

Ask them if they notice a pattern. The correct answer is that prime clock numbers are always Yes's and composite clock numbers are always No's. You might have them verify this hypothesis by doing clock numbers 13, 14, and 15—or all clocks up to 20—depending upon how much practice you want them to have in their multiplication and division skills.

For your own information, prime number clocks are excellent concrete examples of both mathematical groups from group theory and modular systems from number theory.

## 29. SQUARE ROOTS & SUM OF DIGITS (D)

Take any two-digit perfect square, such as 25, and add up the digits, thereby getting $2 + 5 = 7$. Since the square root of 25 is obviously 5, these two numbers are not the same.

Ask your students to find a two-digit perfect square such that the sum of its digits is equal to its square root. This is good practice in obtaining both perfect squares and square roots. It is a fairly simple problem and should not take very long for someone to come up with $81 = 9^2$ and $8 + 1 = 9$.

## 30. SUMS OF TWIN PRIMES (F)

Recall that the Goldbach Conjecture states that every even number (other than 2) can be written as the sum of two prime numbers. We will now modify the conjecture and state a similar type problem for "twin primes".

Twin primes are defined as pairs of prime numbers that are consecutive odd numbers. Some examples are 11 and 13, 17 and 19, 29 and 31, *etc.* Can every even number (other than 2) be written as the sum of two prime numbers, each of which is part of a twin prime pair?

It is immediately noticeable that 4 cannot be written in this way, so if we agree to exclude both 2 and 4, can every other even number be written as the sum of two twin prime numbers?

Let's try a two-digit even number, such as 74, which can be written as the sum of 71 and 3. 3 is a twin prime since 5 is prime, and 71 is a twin prime since 73 is prime.

However, it turns out that not every two-digit even number can be written as the sum of two twin primes. There are three two-digit numbers for which this fails to happen.

Ask your students to try to find one of them. It will give them good practice in both their division and their ability to stick to a problem. The three numbers are 94, 96, and 98.

## 31. COMPOSITES AND MULTIPLES OF 6 (G)

This problem is most easily described using a bit of algebraic notation, but it can also be described with only arithmetic. Although it is true that every prime number—with the exception of 2 and 3—is of the form $6 \times n - 1$ or $6 \times n + 1$, we have seen that it is not true that every number of the form $6 \times n - 1$ or $6 \times n + 1$ is a prime (see Problem #27).

Certainly these numbers are odd, but a number like $6 \times 4 + 1 = 25$ is not prime, and a number like $6 \times 6 - 1 = 35$ is not prime. However, it is rather unusual for both $6 \times n - 1$ and $6 \times n + 1$ to not be prime, for the same number n.

For example, if $n = 5$, $6 \times 5 - 1 = 29$ and $6 \times 5 + 1 = 31$, and both 29 and 31 are prime numbers. If $n = 8$, $6 \times 8 - 1 = 47$ and $6 \times 8 + 1 = 49$, and 47 is a prime number. If $n = 13$, $6 \times 13 - 1 = 77$ and $6 \times 13 + 1 = 79$, and 79 is a prime number.

Ask your students to find a number n such that both $6 \times n - 1$ and $6 \times n + 1$ are not prime. This will take a bit of work, but someone will probably come up with $n = 20$, as $6 \times 20 - 1 = 119 = 17 \times 7$ and $6 \times 20 + 1 = 121 = 11 \times 11$.

Without using algebra, the problem can be presented in the context of finding a multiple of 6 such that both subtracting and adding 1 to this multiple of 6 results in a composite number.

## 32. FAREY FRACTIONS (H, J)

### Stage 1

Define a Farey sequence to be a sequence of fractions in the following way. As a first example, a Farey sequence of order 3 is the sequence of the following fractions: 0/1, 1/3, 1/2, 2/3, and 1/1. Notice that this sequence of fractions consists of all fractions from 0 to 1 with a denominator less than or equal to 3.

Each fraction is reduced to lowest terms and listed in increasing order of size. The Farey sequences of a given order up to 5 are as follows (Table **31**):

**Table 31. Stage 1 of Farey fractions problem.**

| |
|---|
| Order 1: 0/1, 1/1 |
| Order 2: 0/1, 1/2, 1/1 |
| Order 3: 0/1, 1/3, 1/2, 2/3, 1/1 |
| Order 4: 0/1, 1/4, 1/3, 1/2, 2/3, 3/4, 1/1 |
| Order 5: 0/1, 1/5, 1/4, 1/3, 2/5, 1/2, 3/5, 2/3, 3/4, 4/5, 1/1 |

A good exercise is to list the Farey fractions of the next three orders: 6, 7 and 8. In the process of doing this your students will get useful practice in both reducing fractions to lowest terms, and in comparing the size of the fractions by finding their least common denominator.

For example, in the Farey sequence of order 6, they have to compare the fractions 2/5 and 1/3 by the equivalent fractions 2/5 = 6/15 and 1/3 = 5/15. However, the most interesting feature of Farey fractions is to look at any two consecutive Farey fractions in any order.

For example, we could use 2/5 and 1/2 in order 5, 2/3 and 3/4 in order 4, 0/1 and 1/3 in order 3, 1/2 and 1/1 in order 2, *etc*. Ask your students to determine what it is that all consecutive Farey fractions in any order have in common.

If they draw a blank, remind them of the way they can determine that two fractions are equivalent; *i.e.*, by cross-multiplying and getting the same number.

For example, 4/6 = 2/3 and 4 x 3 = 6 x 2. When any two consecutive Farey fractions are cross-multiplied (beginning with the numerator of the first fraction) their difference is always 1. Thus for 2/5 and 1/2, 5 x 1 − 2 x 2 = 1; for 2/3 and 3/4, 3 x 3 − 4 x 2 = 1; for 0/1 and 1/3, 1 x 1 − 0x 3 = 1; and for 1/2 and 1/1, 2 x 1 − 1 x 1 = 1. This is the defining characteristic of successive Farey fractions.

**Stage 2**

Let's once again look at the Farey fractions sequence of order 5: 0/1, 1/5, 1/4, 1/3, 2/5, 1/2, 3/5, 2/3, 3/4, 4/5, 1/1.

Take any three successive Farey fractions in this sequence and add the numerators and denominators of the first and last fractions. For example, using 1/3, 2/5, and 1/2, we would get 1 + 1 = 2 for the numerator and 3 + 2 = 5 for the denominator, thereby making the fraction 2/5—which is exactly the middle fraction. Does this phenomenon always occur?

Have your students try a number of examples, using the Farey sequences of the orders they have already constructed. Notice that for 2/3, 3/4, and 4/5 we get 2 + 4 = 6 and 3 + 5 = 8; but 6/8 = 3/4 when reduced to lowest terms. There are no surprises in this particular exercise, as this phenomenon always does exist in Farey fractions of any given order.

## 33. SUMS OF FACTORIALS OF DIGITS (D)

By the factorial of a number we mean the product of the number with all consecutive numbers less than it—down to 1. For example, 5! means: 5 x 4 x 3 x 2 x 1 = 120.

Take the factorial of each digit in a three-digit number, then add these factorials together. For example, using the number 153, we would have: 1! = 1, 5! = 20, and 3! = 6.

Their sum is 127, which is not equal to the original number, 153.

Find a three-digit number such that the sum of the factorials of its digits equals the number. The problem requires some persistence. Hint: The number is between 100 and 200 (or 100 and 150). The answer is 145, as 1! = 1, 4! = 24, 5! = 120 and 120 + 21 + 1 = 145.

## 34. SUMS OF CUBES OF DIGITS (D)

This problem is similar in context to the sum of factorials of digits problem. Take a three-digit number such as 145 and write down the cubes of each digit: $1^3 = 1$, $4^3 = 64$, $5^3 = 125$.

Add these three cubes, 1 + 64 + 125 = 190, which does not equal our original number, 145.

The problem is to find a three-digit number such that the sum of the cubes of its digits equals the original number. Hint: The number is again between 100 and 200.

The answer is 153; $1^3 = 1$, $5^3 = 125$, $3^3 = 27$, and 1 + 125 + 27 = 153.

## 35. SUMS OF THREE CUBES (D)

The problem is to find a cube that is the sum of three cubes. For example, $4^3 = 64$, and if we try to write 64 as the sum of three cubes, even allowing for using a cube more than once, a little trial and error will show that it cannot be done. Notice that we are limited to the cubes $1^3 = 1$, $2^3 = 8$, and $3^3 = 27$.

Hint: The cube we are looking for (*i.e.*, the cube that can be written as the sum of three cubes), comes from a single-digit number.

The answer is $6^3 = 216$; $216 = 3^3 + 4^3 + 5^3$ as $3^3 = 27$, $4^3 = 64$, $5^3 = 125$, and $27 + 64 + 125 = 216$.

## 36. LINEAR DIOPHANTINE EQUATIONS (I, J)

We come now to a cornerstone of number theory: Diophantine equations. This is generally appreciated most with a little background in algebra, although it can be given to good pre-algebra students who are familiar with signed numbers, as a motivation device to begin the study of algebra.

### Stage 1

Any equation in two variables can always be solved in general, such as $2x + 3y = 6$ can be solved as $3y = -2x + 6$ and $y = -2x/3 + 2$. Notice that $x = 0$ and $y = 2$ is a solution of the original equation, as is $x = 3/2$ and $y = 1$. The first solution, $x = 0$ and $y = 2$, involves only integers (*i.e.*, positive and negative counting numbers).

When we are interested in solving an equation in integers only—as opposed to fractions or decimals—we call it a Diophantine equation. If it is of the first degree (*i.e.*, it has no exponents greater than 1), we call it a Linear Diophantine equation. Thus, we see that $2x + 3y = 6$ has a solution in integers and therefore can be solved as a Linear Diophantine equation.

The same statement holds for the equation $3x - 6y = 15$, as $x = -1$ and $y = -3$ is a solution to this equation in integers. However, it appears that the equation $2x + 4y = 3$ does not have a solution in integers. Begin making the following table as we add more equations (Table **32**).

Table 32. Stage 1 of linear Diophantine equations problem.

| Yes | No |
|---|---|
| 2x + 3y = 6<br>2x + 4y = 3 | |
| 3x − 6y = 15 | 5x − 10y = 4 |

*(Table 32) contd.....*

| 3x − 4y = 1 | 6x + 8y = 5 |
|:---:|:---:|

The question is: How do we know when we are able to solve a Linear Diophantine equation in integers?

After the students compose a number of examples for which equations can be solved and not solved, they will see certain characteristics. Perhaps they will notice that when the numbers next to the variables are relatively prime, the equation can always be solved: some examples are 2x + 3y = 6 and 3x + 4y = 1. Perhaps they will notice that when the numbers next to both variables are even and the number by itself is odd, the equation cannot be solved.

It is a difficult problem to solve completely, but any partial solution should be considered as good work and effort. The answer is that Linear Diophantine equations can be solved if and only if the number by itself is a multiple of the greatest common divisor of the numbers next to the variables.

For example, in 2x + 3y = 6 the greatest common divisor for 2 and 3 is 1; therefore any number by itself will work. But for 5x − 10y = 4, the greatest common divisor of 5 and 10 is 5, 4 is not a multiple of 5, and the equation can therefore not be solved in integers. For a more sophisticated example, you might suggest that the children try 18x + 27y = 33, which cannot be solved in integers since the greatest common divisor of 18 and 27 is 9, and 33 is not a multiple of 9, even though 33 is a multiple of 3, and 3 is a divisor of both 18 and 27.

## Stage 2

An additional feature of looking for solutions to Linear Diophantine equations involves an application to Analytic Geometry. When students are first learning the Cartesian Coordinate system and graphing first degree equations as straight lines, a playful application of Linear Diophantine equations can be presented. The application involves asking if a straight line will always eventually pass through a "lattice point" on the graph—which means a point where both pairs of numbers are integers. Look at the graph 2x + 3y = 13. Make a table of values, such as the following (Table **33**):

**Table 33. Stage 2a of linear Diophantine equations problem.**

| X | Y |
|:---:|:---:|
| 0 | 13/3 |
| 1 | 11/3 |
| 2 | 3 |

We can see from the above table that the line passes through the lattice point (2,3).

This is equivalent to saying that the Linear Diophantine equation $2x + 3y = 13$ can be solved with the integers $x = 2$ and $y = 3$. However, if we look at the graph of $2x + 4y = 7$, a possible table of values is (Table **34**):

**Table 34. Stage 2b of linear Diophantine equations problem.**

| X | Y |
|---|---|
| 0 | 7/4 |
| 1 | 5/4 |
| 2 | 3/4 |

A good question is: Will the above line *ever* pass through a lattice point (*i.e.*, a pair of integers)? Stage 1 shows that it will never happen because 7 is not a multiple of the greatest common divisor of 2 and 4, which is 2. Therefore, this Diophantine equation has no solution in integers, which is equivalent to saying that the line represented by the equation $2x + 4y = 7$ will never pass through a lattice point. This is an interesting interplay between Linear Diophantine equations and Analytic Geometry.

## 37. PAIRS OF SQUARES (D)

Take a two-digit number and square it. For example $71^2 = 5,461$, or $45^2 = 2,025$. When we divide the numbers 5,461 in pairs, we get 54 and 61—neither one of which is a perfect square. When we divide the numbers 2,025 in pairs, 25 is a perfect square but 20 is not a perfect square. If we happen to get only 3 digits, such as $31^2 = 961$, we consider the numbers to be 0961, and $09 = 9$ is a perfect square but 61 is not a perfect square. The problem is to find a perfect square such that each pair of digits is also a perfect square. After a bit of trial and error, the forthcoming answer is that $41^2 = 1,681$, and 16 and 81 are both perfect squares; 16 $= 4^2$ and $81 = 9^2$.

## 38. SQUARES AND CONSECUTIVE DIGITS (D)

This is similar to the above problem. Look for a perfect square whose pairs of digits are consecutive numbers. For example, $71^2 = 5461$, and 54 and 61 are not consecutive numbers; $55^2 = 3,025$, and 30 and 25 are not consecutive numbers; $31^2 = 963$, and 9 and 63 are not consecutive numbers. The answer is $91^2 = 8,281$, as 81 and 82 are indeed consecutive numbers.

## 39. POWERS OF 11 AND PALINDROMES (D)

A palindrome is a number such that when its digits are reversed, the same number results. For example, 77, 232, and 12321 are palindromes.

An interesting phenomenon results when we look at the powers of 11: $11^1 = 11$, $11^2 = 121$, $11^3 = 1331$. Notice that each of these powers of 11 are palindromes. Are all powers of 11 palindromes?

Continuing the procedure, your students will find that $11^4 = 14641$ is again a palindrome. However, it should not take them very long to determine that not all powers of 11 are palindromes, as $11^5 = 161651$ is not a palindrome.

## 40. GREATEST COMMON DIVISOR AND LEAST COMMON MULTIPLE (G, H)

This next problem is great for children who are learning fractions.

**Stage 1**

Let's practice finding the Greatest Common Divisor (GCD) and Least Common Multiple (LCM) using prime factorization.

A problem of this type comes up naturally when adding fractions, such as 3/42 + 8/9. In order to find the LCM of 42 and 9, always take multiples of 42 until 9 divides into it; 42, 84, and 126, for example. (9 divides into 126 evenly; therefore, 126 is the LCM for 42 and 9).

However, a more interesting way of doing this, and one that is quite useful for large numbers, involves writing down the prime factorization for each number: 42 = 2 x 3 x 7; $9 = 3^2$. Simply list each prime that occurs the most number of times it occurs in any given expression; then multiply.

The answer is $2 \times 3^2 \times 7 = 2 \times 9 \times 7 = 126$. This process is extremely useful when we have more than two fractions to work with. For example, add 1/3 + 3/8 + 5/12 + 4/15. The LCM is therefore $2^3 \times 3 \times 5 = 120$ since $8 = 2^3$, $12 = 2^2 \times 3$ and 15 = 5 x 3.

A similar procedure can be used to find the GCD of two or more numbers. Use 42 and 9 again. The prime factorizations of $42 = 2 \times 3^2 \times 7$ and $9 = 3^2$.

Use only primes that are in both factorizations (*i.e.*, the least number of times they occur in any given expression). The result is simply 3, and 3 is the GCD of 42 and 9.

Using 80 and 120, $80 = 2^4 \times 5$ and $120 = 2^3 \times 3 \times 5$; the GCD is $2^3 \times 5 = 40$ and the LCM is $2^4 \times 3 \times 5 = 240$.

## Stage 2

A pleasing relationship exists between the LCM and GCD. Use the previous example of 42 and 9. The LCM was 126 and the GCD was 3. Multiply the LCM and GCD together. The answer is $126 \times 3 = 378$. Now multiply the original two numbers together. The answer is $42 \times 9 = 378$ too. Does this phenomenon always hold?

Ask your students to try their own set of numbers and compose a table. Notice that if two numbers are relatively prime, meaning that no number other than 1 divides into both of them, their GCD is 1 and their LCM is the product of the two numbers.

For example, 3 and 7 are relatively prime. Their GCD = 1 and their LCM = 21, and $3 \times 7 = 1 \times 21 = 21$. After some experimentation, it will appear that this formula does always seem to work. The reason is that the primes not used in the LCM are exactly the primes used in the GCD—thereby forming the unique factorization of the two numbers.

## 41. FACTORIAL-PRIME DIVISIBILITY CRITERIA (G, J)

This problem displays a most interesting relationship among factorials, primes, and divisibility. Take a one-digit number, such as 5, subtract 1—yielding 4, and form $4! = 4 \times 3 \times 2 \times 1 = 24$. Then add 1, yielding 25, and notice that the original number, 5, divides into the final result, 25, evenly.

Working through this procedure for another number, such as 6, yields $51 = 5 \times 4 \times 3 \times 2 \times 1 = 120$ and notice that $120 + 1 = 121$ is not divisible by 6.

Form a table from 2 to 11 (or higher depending upon the division ability of your class). A familiar pattern will follow, as prime numbers have the distinction of always being divisible into the resulting number obtained by this process, whereas composite numbers always end up not being divisible into the resulting number.

This is actually a way of characterizing prime numbers. Algebraically, the formula can be stated as $(p - 1)! + 1$ is divisible by p if and only if p is a prime number.

## 42. FACTORIAL-COMPOSITE DIVISIBILITY CRITERIA (G, J)

As a change of pace from dealing so much with prime numbers, we will now look at a phenomenon that works for practically all composite numbers but does not work for prime numbers. Take a single-digit number, such as 6, subtract 1, yielding 5, and form the factorial $5! = 5 \times 4 \times 3 \times 2 \times 1 = 120$.

Now notice that the original number 6 divides evenly into the resulting number 120.

Try the same procedure for 7 and notice that 7 does not divide evenly into $6! = 720$. Work through this procedure for all numbers from 2 through 11 (or higher depending upon the divisibility ability of your class).

With one exception, this phenomenon works for every composite number and does not work for any prime number. The exception is that it does not work for the composite number 4, as $3! = 3 \times 2 \times 1 = 6$ and 4 does not divide evenly into 6.

Algebraically this formula can be stated as follows: for every composite number n $\neq 4$, $(n-1)!$ is divisible by n.

## 43. SUMS OF DIVISORS AND SQUARES (G)

A nice little problem that combines the operations of addition, single-digit multiplication, and single-digit division, is as follows. Find a number such that the sum of all the divisors of the number is a perfect square.

The smallest number that works is 3, as its divisors are 1 and 3, and $1 + 3 = 2^2$. The next number that works is 22; the divisors of 22 are 1, 2, 11, and 22, and $1 + 2 + 11 + 22 = 36 = 6^2$.

Have your students try to come up with the three remaining two-digit numbers that have this property. The numbers are 66, 70, and 81.

## 44. MORE ON FIBONACCI NUMBERS (D, G, J)

**Stage 1**

Another relationship within the Fibonacci numbers can be seen by again putting the Fibonacci numbers in the first column, but this time putting the squares of the Fibonacci numbers in the second column, as can be seen in Table **35**.

This relationship is by no means obvious, and a good hint will most likely be necessary. Tell your students to try multiplying alternate Fibonacci numbers together and see what develops, such as $1 \times 2 = 2$, $1 \times 3 = 3$, $2 \times 5 = 10$, $3 \times 8 = 24$, $5 \times 13 = 65$, $8 \times 21 = 168$, $13 \times 34 = 442$, *etc*.

**Table 35. Stage 1a of more on Fibonacci numbers problem.**

| Fibonacci Numbers | Squares of Fibonacci Numbers |
|:---:|:---:|
| 1 | 1 |
| 1 | 1 |
| 2 | 4 |
| 3 | 9 |
| 5 | 25 |
| 8 | 64 |
| 13 | 169 |
| 21 | 441 |
| 34 | 1156 |
| 55 | 3025 |
| 89 | 7921 |

Writing down the numbers in a third column, the table now looks like the following (Table **36**; F numbers mean Fibonacci numbers):

**Table 36. Stage 1b of more on Fibonacci numbers problem.**

| F Numbers | Squares of F Numbers | Products of Alternate F Numbers |
|:---:|:---:|:---:|
| 1 | 1 | |
| 1 | 1 | 2 |
| 2 | 4 | 3 |
| 3 | 9 | 10 |
| 5 | 25 | 24 |
| 8 | 64 | 65 |
| 13 | 169 | 168 |
| 21 | 441 | 442 |
| 34 | 1156 | 1155 |
| 55 | 3025 | 3026 |
| 89 | 7921 | 7920 |

The basic relationship should be fairly clear, although it is still rather puzzling.

Trying a few more numbers will be good multiplication practice and lots of fun.

The numbers in the third column apparently are, alternatively, 1 more than and 1 less than the numbers in the second column: $1 + 1 = 2$, $4 - 1 = 3$, $9 + 1 = 10$, $25 - 1 = 24$, $64 + 1 = 65$, $169 - 1 = 168$, $441 + 1 = 442$, *etc.*

Algebraically this formula can be stated, using the fact that $(-1)^n = +1$ if n is even and $(-1)^n = -1$ if n is odd, as $F_{n-1} \times F_{n+1} - F2 = (-1)n^2 = (-1)^n$.

## Stage 2

Fibonacci numbers are quite rich in their exploration and discovery possibilities. Yet another pattern emerges by again making a table like the first two columns in Stage 1, with the Fibonacci numbers in the first column and their squares in the second column.

This time you might try giving your students the go-ahead in finding out the pattern all by themselves.

If a hint seems necessary, you can tell them to try multiplying successive Fibonacci numbers together: $1 \times 1 = 1$, $1 \times 2 = 2$, $2 \times 3 = 6$, $3 \times 5 = 15$, $5 \times 8 = 40$, $8 \times 13 = 104$, *etc* (Table **37**):

Table 37. Stage 2a of more on Fibonacci numbers problem.

| Fibonacci Numbers | Squares of Fibonacci Numbers |
|:---:|:---:|
| 1 | 1 |
| 1 | 1 |
| 2 | 4 |
| 3 | 9 |
| 5 | 25 |
| 8 | 64 |
| 13 | 169 |
| 21 | 441 |
| 34 | 1156 |

Let's add this third column to our list (Table **38**):

**Table 38. Stage 2b of more on Fibonacci numbers problem.**

| F Numbers | Squares of F Numbers | Products of Successive F Numbers |
|:---:|:---:|:---:|
| 1 | 1 | 1 |
| 1 | 1 | 2 |
| 2 | 4 | 6 |
| 3 | 9 | 15 |
| 5 | 25 | 40 |
| 8 | 64 | 104 |
| 13 | 169 | 273 |
| 21 | 441 | 714 |
| 34 | 1156 | 1870 |

It should be evident that the numbers in the third column, added to the numbers in the second column in the row below it, yield the numbers in the row below the original numbers in the third column.

For example, $1 + 1 = 2$, $2 + 4 = 6$, $6 + 9 = 15$, $15 + 25 = 40$, $40 + 64 = 104$, $104 + 169 = 273$, *etc.*

Perhaps someone will notice that the sum of the numbers in the second column always equals the corresponding number in the third column: $1 = 1$, $1 + 1 = 2$, $1 + 1 + 4 = 6$, $1 + 1 + 4 + 9 = 15$, $1 + 1 + 4 + 9 + 25 = 40$, $1 + 1 + 4 + 9 + 25 + 64 = 104$, $1 + 1 + 4 + 9 + 25 + 64 + 169 = 273$, *etc.*

These characteristics are two equivalent ways of describing the formula that can be algebraically stated as

$$\sum_{k=1}^{N} F_{k2} = F_k \cdot F_{k+1}$$

**Stage 3**

We now enter the territory of exploring the relationship of Fibonacci numbers and divisibility.

Let's construct the first 15 numbers in the Fibonacci sequence; we'll form two columns, listing the numbers 1 through 15 in the first column, and their corresponding Fibonacci numbers in the second column (Table **39**):

**Table 39. Stage 3 of more on Fibonacci numbers problem.**

| Counting Numbers | Fibonacci Numbers |
|:---:|:---:|
| 1 | 1 |
| 2 | 1 |
| 2 | 2 |
| 4 | 3 |
| 5 | 5 |
| 6 | 8 |
| 7 | 13 |
| 8 | 21 |
| 9 | 34 |
| 10 | 55 |
| 11 | 89 |
| 12 | 144 |
| 13 | 233 |
| 14 | 377 |
| 15 | 610 |

Let's see which numbers divide evenly into the numbers in the first column. 1 divides into all the numbers, of course; 2 divides into 2, 4, 6, 8, 10, 12, and 14; 3 divides into 3, 6, 9, 12, and 15; 4 divides into 4, 8 and 12; 5 divides into 5, 10 and 15; 6 divides into 6 and 12; 7 divides into 7 and 14.

Let's check the situation for the corresponding Fibonacci numbers. The first Fibonacci number is 1, and this divides evenly into all the Fibonacci numbers.

Similarly, the second Fibonacci number is also 1 and thereby divides into all Fibonacci numbers. The third Fibonacci number is 2, and 2 divides into 2, 8, 34, 144, and 610; these are the third, sixth, ninth, twelfth, and fifteenth Fibonacci numbers. 2 does not divide into any of the other Fibonacci numbers in the second column.

Have your students determine if this amazing correspondence always takes place for these first 15 Fibonacci numbers.

This correspondence always works, and we can state it algebraically by saying: *If m divides n (where m and n are positive integers) then $F_m$ divides $F_n$ (where $F_m$ and $F_n$ are the mth and nth Fibonacci numbers).*

For Fibonacci enthusiasts we can go further. Notice that if the Greatest Common Divisor (GCD) of p and q is r, then the GCD of $F_p$ and $F_q$ is $F_r$.

For example, the GCD of 8 and 12 is 4, and the GCD of $F_8 = 21$ and $F_{12} = 144$ is $F_4 = 3$.

## Stage 4

A final problem that can be done with the Fibonacci numbers has to do with once again examining the relationship of the Fibonacci sequence to the Lucas sequence.

Recall that the Lucas sequence is the numbers in the sequence 1, 3, 4, 7, 11, 18, 29, 47, *etc*. We have seen that the sum of alternate Fibonacci numbers equals the corresponding Lucas number; algebraically this was stated as

$$L_n = F_{n-1} + F_{n+1}.$$

Let's list the Fibonacci numbers and Lucas numbers in two columns (Table **40**):

Table 40. Stage 4a of more on Fibonacci numbers problem.

| Fibonacci Numbers | Lucas Numbers |
|---|---|
| 1 | 1 |
| 1 | 3 |
| 2 | 4 |
| 3 | 7 |
| 5 | 11 |
| 8 | 18 |
| 13 | 29 |
| 21 | 47 |
| 34 | 76 |
| 55 | 123 |
| 89 | 199 |
| 144 | 322 |
| 233 | 521 |
| 377 | 843 |

As we already observed, we have $1 + 2 = 3$, $1 + 3 = 4$, $2 + 5 = 7$, $3 + 8 = 11$, $5 + 13 = 18$, $8 + 21 = 29$, *etc*. Another relationship can be seen by multiplying the corresponding Fibonacci numbers and Lucas numbers together, forming a third column: (Table **41**; "F numbers" mean Fibonacci numbers and "L numbers" mean Lucas numbers):

**Table 41. Stage 4b of more on Fibonacci numbers problem.**

| F Numbers | L Numbers | Products of F Numbers & L Numbers |
|:---:|:---:|:---:|
| 1 | 1 | 1 |
| 1 | 3 | 3 |
| 2 | 4 | 8 |
| 3 | 7 | 21 |
| 5 | 11 | 55 |
| 8 | 18 | 144 |
| 13 | 29 | 377 |

The third column is, once again, Fibonacci numbers. However, they are very particular Fibonacci numbers.

Considering 1 to be the second Fibonacci number, we have the second, fourth, sixth, eighth, tenth, twelfth, and fourteenth Fibonacci numbers.

When we multiply a Fibonacci number by its corresponding Lucas number, we double the order of the Fibonacci number. Algebraically this formula can be stated as $F_n \times L_n = F_{2n}$. We thus end the amazing world of the Fibonacci numbers.

## 45. SUMS OF SQUARES AS A SQUARE (D)

This is a good problem to practice finding square roots through estimation by multiplication Let's take the sum of the first few squares and check to see if we have a perfect square when we add them up: $1^2 = 1$, $1^2 + 2^2 = 5$, $1^2 + 2^2 + 3^2 = 14$, $1^2 + 2^2 + 3^2 + 4^2 = 32$, $1^2 + 2^2 + 3^2 + 4^2 + 5^2 = 55$, *etc.*

Have your students continue this process until they find a sum that is a perfect square. When they reach large numbers (which of course they will), they will have to use estimation to determine if the number is a perfect square.

For example, if they were checking to see if 1,000 were a perfect square, they could try a number such as 35; $35^2 = 1,225$. Since 35 is too large, they might try $32^2 = 1,024$; since 32 is still a little too large, they might try $31^2 = 961$ and thereby determine that 1,000 is not a perfect square.

With persistence and accuracy, they will eventually come up with the fact that the smallest number of consecutive perfect squares they must add to reach a perfect square is 24: $1^2 + 2^2 + 3^2 + ... + 24^2 = 4,900 = 70^2$.

## 46. FACTORIAL MULTIPLICATION (D,J)

This is a brief but interesting little problem that practices multiplication of large numbers in the context of working with factorials.

Recall that 5! = 5 x 4 x 3 x 2 x 1 = 120, 61 = 6 x 5 x 4 x 3 x 2 x 1 = 720, 71 = 7 x 6 x 5 x 4 x 3 x 2 x 1 = 5040, *etc.*

Ask your students to first compile a list of all the factorials from 1 to 10; *i.e.*, 1!, 2!, 3! ... up to 10!. Then ask them to find two numbers such that when their factorials are multiplied together they yield the number that is 10! Other than 1 and 10, the correct numbers are 6 and 7 as 6! x 7! = 10!.

## 47. SUMS OF PRIMES AND DIVISORS (F)

In this problem we will first need to find all the prime numbers less than a given number.

For example, if our given number is 30, the prime numbers less than 30 are 2, 3, 5, 7, 11, 13, 17, 19, 23, and 29. The sum of all these prime numbers is 129, and clearly 30 does not divide evenly into 139.

A single-digit number that does have this property is 5. The prime numbers less than 5 are 2 and 3; 2 + 3 = 5, and 5 certainly divides evenly into 5.

There is one and only one two-digit number that also has this property, and it will be the job of your students to find this number. This problem will take a bit of time, but is excellent practice in division, persistence, accuracy, and finding prime numbers.

The correct answer is 71; the sum of all the prime numbers less than 71 is 568 and 71 divides evenly into 568.

## 48. PRIME REVERSALS (G)

In this problem we look for prime numbers that have the interesting feature that when their digits are reversed, the numbers are still prime.

You can start with two-digit numbers; some examples are 11, 13, and 17; 11, 31, 71 are the numbers obtained when their digits are reversed, and these numbers are all prime.

There are two other two-digit numbers that have this property, and if your students make a list of all the two-digit prime numbers, they should not have too much trouble figuring out that the remaining two-digit prime numbers with this reversal property are 37 and 79, as 73 and 97 are also prime.

The problem becomes more challenging when we allow three-digit prime numbers to enter the picture. Two three-digit prime numbers with this prime reversal property are 113 and 337, as their reversals are 311 and 733 and it can be verified that these numbers are both prime. There is one more three-digit prime number with this property, and a good persevering student who has become adept in the area of Recreational Number Theory may be able to find it.

Hint: the number is between 100 and 200. The answer is 199, as 991 is also a prime number.

## 49. MORE ON TRIANGULAR NUMBERS (D, G, J)

Just as we developed a number of additional interesting patterns and formulas for Fibonacci numbers, we will now explore some more of the richness that exists in the world of Triangular numbers.

### Stage 1

Recall that the Triangular numbers are the numbers in the sequence 1, 3, 6, 10, 15, 21, 28, 36, 45, 55, 66, *etc.* A series of three relatively easy problems with Triangular numbers is as follows:

**Problem A:** Find a pair of successive Triangular numbers whose sum and difference is also Triangular. For example, 3 and 6 are successive Triangular numbers whose difference, $6 - 3 = 3$, is a Triangular number but whose sum, $3 + 6 = 9$, is not a Triangular number.

The correct answer is 15 and 21, as $15 + 21 = 36$ and $21 - 15 = 6$ are both Triangular numbers.

(A more challenging answer is the sum of Triangular numbers 780 and 990 as $780 + 990 = 1,170$ and $990 - 780 = 210$ are both Triangular numbers.)

**Problem B:** Find a Triangular number, other than 1, whose square is also triangular. This is easy, as $3^2 = 9$ is not triangular—but $6^2 = 36$ is triangular.

**Problem C:** Find a three-digit palindromic Triangular number. Recall that a palindrome is a number which yields the same number when its digits are reversed, such as 232.

This problem might take a bit more work, but is still easy. Just continue the Triangular numbers for a while: 1, 3, 6, 10, 15, 21, 28, 36, 45, 55, 66, 78, 91, 105, 120, 136, 153, and finally 171 is a three-digit palindromic Triangular number.

Two other three-digit palindrome triangular numbers are 595 and 666. Palindrome enthusiasts may find a four-digit palindrome Triangular number: 5,995, and even a five-digit palindrome Triangular number: 20,302.

## Stage 2

In our previous work on Triangular numbers, we have seen that the nth Triangular number can be found as $n(n+1)/2$, each perfect square is the sum of two successive Triangular numbers, and the sum of the first n cubes is the square of the nth Triangular number.

Another interesting property of Triangular numbers is related to consecutive odd squares. Compose the following table with Triangular numbers in the first column and consecutive odd squares in the second column (Table **42**):

Table 42. Stage 2 of more on triangular numbers problem.

| Triangular Numbers | Consecutive Odd Squares |
|:---:|:---:|
| 1 | 9 |
| 3 | 25 |
| 6 | 49 |
| 10 | 81 |
| 15 | 121 |
| 21 | 169 |
| 28 | 225 |
| 36 | 289 |
| 45 | 361 |
| 55 | 441 |
| 66 | 529 |
| 78 | 729 |
| 91 | 841 |
| 105 | 1961 |

Ask your students to determine the relationship between the Triangular numbers and the consecutive odd squares. Hint: Look at the Triangular numbers in the first column and match them to the Consecutive Odd Square numbers in the second column.

What do these pairs of numbers always have in common? For example, 1 and 9, 3 and 25, 6 and 49, 10 and 81, 15 and 121, *etc*. The second number is always eight times the first number, plus one.

This is an interesting pattern: Every odd square can be written as eight times a triangular number, plus 1. Algebraically we can state the exact relationship, where $T_n$ is the nth Triangular number, as $(2n+1)^2 = 8T_n + 1$.

## Stage 3

A somewhat less obvious relationship that exists within Triangular numbers can be seen by writing down the first 18 Triangular numbers in a column (Table **43**):

Table 43. Stage 3 of more on triangular numbers problem.

| Triangular Numbers | |
|---|---|
| $T_1$ | 1 |
| $T_2$ | 3 |
| $T_3$ | 6 |
| $T_4$ | 10 |
| $T_5$ | 15 |
| $T_6$ | 21 |
| $T_7$ | 28 |
| $T_8$ | 36 |
| $T_9$ | 45 |
| $T_{10}$ | 55 |
| $T_{11}$ | 66 |
| $T_{12}$ | 78 |
| $T_{13}$ | 91 |
| $T_{14}$ | 105 |
| $T_{15}$ | 120 |
| $T_{16}$ | 136 |
| $T_{17}$ | 153 |
| $T_{18}$ | 171 |

Now let's take sums of groups of Triangular numbers in the following ways: $T_1 +$

$T_2 + T_3 = 1 + 3 + 6 = 10 = T_4$; $T_5 + T_6 + T_7 + T_8 = 15 + 21 + 28 + 36 = 100 = 45 + 55 = T_9 + T_{10}$; $T_{11} + T_{12} + T_{13} + T_{14} + T_{15} = 66 + 78 + 91 + 105 + 120 = 460 = 136 + 153 + 171 = T_{16} + T_{17} + T_{18}$.

Have your students see if this amazing relationship continues by determining if $T_{19} + T_{20} + T_{21} + T_{22} + T_{23} + T_{24} = T_{25} + T_{26} + T_{27} + T_{28}$ (It does). This is a wonderful exercise in addition.

## Stage 4

A more typical relationship that can be done in Triangular numbers that resembles something that was done earlier in Fibonacci numbers involves taking squares of Triangular numbers.

Construct a column of triangular numbers and another column of their corresponding squares (Table **44**):

Table 44. Stage 4a of more on triangular numbers problem.

| Triangular Numbers | Triangular Squares |
|---|---|
| 1 | 1 |
| 3 | 9 |
| 6 | 36 |
| 10 | 100 |
| 15 | 255 |
| 21 | 441 |
| 28 | 784 |
| 36 | 1296 |
| 45 | 2025 |
| 55 | 3025 |

Next, multiply alternate Triangular numbers together, forming a third column (Table **45**):

Table 45. Stage 4b of more on triangular numbers problem.

| Triangular Numbers | Triangular Squares | Alternate Products |
|---|---|---|
| 1 | 1 | 6 |
| 3 | 9 | 30 |

*(Table 45) contd.....*

| | | |
|---|---|---|
| 6 | 36 | 90 |
| 10 | 100 | 210 |
| 15 | 255 | 420 |
| 21 | 441 | 756 |
| 28 | 784 | 1260 |
| 36 | 1296 | 1296 |

Your students should have no trouble in observing that the sum of a number in the first column with the corresponding number in the preceding row of the third column yields the corresponding number in the second column that is in the same row as the original number in the first column: $3 + 6 = 9$, $6 + 30 = 36$, $10 + 90 = 100$, $15 + 210 = 225$, $21 + 420 = 441$, *etc.*

Algebraically, we can state this formula as the following: $T_n^2 = T_n + (T_{n-1})(T_{n+1})$.

## Stage 5

As a final problem with Triangular numbers, we will determine a most unusual relationship between Triangular numbers and perfect squares. Construct a table with the Triangular numbers in the first column and the perfect squares in the second column.

Notice that the first Triangular number $T_1 = 1$, is the sum of the first perfect square by itself: $1 = 1^2$. The second Triangular number, $T_2 = 3$ can be written as $4 - 1$, or $2^2 - 1^2$. $T_3 = 6$ can be written as $9 - 4 + 1$ or $3^2 - 2^2 + 1^2$; $T_4 = 10 = 16 - 9 + 4 - 1$, or $4^2 - 3^2 + 2^2 - 1^2$; $T_5 = 15 = 25 - 16 + 9 - 4 + 1 = 5^2 - 4^2 + 3^2 - 2^2 + 1^2$ (Table **46**):

**Table 46. Stage 5 of more on triangular numbers problem.**

| Triangular Numbers | Perfect Squares |
|---|---|
| 1 | 1 |
| 3 | 4 |
| 6 | 9 |
| 10 | 16 |
| 15 | 25 |
| 21 | 36 |
| 28 | 49 |
| 36 | 64 |

*(Table 46) contd.....*

| 45 | 81 |
|---|---|
| 55 | 100 |

Have your students verify that this unusual pattern works—through $T_{10}$. This is a true formula for all Triangular numbers and perfect squares, mysterious as it may seem.

The algebraic formula can be described as the following: $T = n^2 - (n-1)^2 + (n-2)^2 - (n-3)^2 + (n-4)^2 - (n-5)^2 + \ldots 1^2$.

## 50. CREATIVE DIGIT OPERATIONS (B, C, D)

After finishing with Triangular numbers, we move on to a series of three short problems involving addition, subtraction, multiplication, and taking exponents of digits.

**Problem A:** The problem is to combine the digits 1 to 9 in any consecutive order, using only the operations of addition, subtraction, and single-digit multiplication, but the result must equal 100.

The problem is more difficult than it may appear. It may be helpful to break the problem into two separate problems, one using only addition and single-digit multiplication, the other using only addition and subtraction.

The answer to the first problem is $1 + 2 + 3 + 4 + 5 + 6 + 7 + (8 \times 9) = 100$.

The answer to the second problem is $123 - 45 - 67 + 89 = 100$.

Your students may or may not solve this problem, but they will get a lot of practice with basic arithmetic in a creative fashion.

**Problem B:** The object of this problem is to find a three-digit number that is a power of 2, each of its digits also being a power of 2.

Remind your students that any number to the zero power is 1, and any number to the 1 power is itself; thus,

$2^0 = 1$ and $2^1 = 2$.

If your students are going about it systematically, they should not take long to solve the problem, as the first three-digit power of 2 they will come to is 128, and $1 = 2^0$, $2 = 2^1$ and $8 = 2^3$.

This is an easy change of pace for the students, but also good, playful introductory practice with exponents.

**Problem C:** This problem is great for practicing multiplication skills. Ask your students to multiply the digits together of the number 73,942 in any order they like, keeping the digits arranged in the same order.

For example, 7 x 3 x 9 x 4 x 2 = 1,512, 73 x 9 x 42 = 27,594, 7 x (3 x 9) x 42 = 7,438, *etc*.

The object is to find two different ways of multiplying the digits together in consecutive order that yield the same number. The two different ways are 73 x 9 x 42 = 27,594 and 7 x 3942 = 27,594.

## 51. NUMBER REVERSALS AND SUBTRACTION (B)

This is an excellent way to practice subtracting four-digit numbers. The problem is to find a four-digit number such that when its reversal is subtracted from it, the digits are left unchanged, though in a different order.

Hint: The number is in the 20th century. For example, the reversal of 1,951 is 1591 and 1951 − 1591 = 360 which, of course, does not leave the digits unchanged.

Notice that if we tried any number with a digit bigger than 1 on the right, we would not be able to subtract its reversal—because the reversal would be bigger than the actual number.

With this in mind, it should not take long for your students to arrive at 1,980, as 1980 − 891 = 1089.

## 52. CYCLES OF CUBES (C)

This problem will require some persistence on the part of your students. The problem is to find two three-digit numbers such that when the cubes of each of the digits of a number are added together, they yield the other number.

This does have a similarity in flavor to pairs of amicable numbers. Hint: One number is between 100 and 200, and the other number is between 200 and 300.

For example, let's try 245: $2^3 = 8$, $4^3 = 64$, $5^3 = 125$, and 125 + 64 + 8 = 197. However, $1^3 = 1$, $9^3 = 729$, $7^3 = 343$, and their sum is 1,073.

This is a challenging problem, but if your students have been practicing Recreational Number Theory for a while, someone may very well be able to come up with the elusive pair of numbers 136 and 244, as $1^3 = 1$, $3^3 = 27$, $6^3 = 216$, and 1 + 27 + 216 = 244; $2^3 = 8$, $4^3 = 64$, $4^3 = 64$ and 8 + 64 + 64 = 136.

## 53. PRODUCTS OF DIVISORS AND SQUARES (G)

Now find a number such that the product of all the proper divisors of the number equals the square of the number. For example, if we tried the number 18, we would have $1 \times 2 \times 3 \times 6 \times 9 = 324 = 18^2$.

The number 12 also works, as $1 \times 2 \times 3 \times 4 \times 6 = 144 = 12^2$. It should not be too difficult for your students to come up with two other two-digit numbers that also work: 20 and 45 are good choices, as $1 \times 2 \times 4 \times 5 \times 10 = 400 = 20^2$ and $1 \times 3 \times 5 \times 9 \times 15 = 2,025 = 45^2$.

## 54. FUN AND GAMES WITH 9 (C)

Next, let's try an exercise that involves nothing more than single-digit multiplication. Take a three-digit number, double it, multiply it by 6, add the digits, multiply it by 3, add the digits, and note the result.

Let's try it for the number 216. Doubling gives 432, multiplying by 6 yields 2,592 (of course we could have just multiplied by 12, but this way we do not have to worry about two-digit multiplication). Adding the digits gives us $2 + 5 + 9 + 2 = 18$, multiplying by 3 yields 54, and adding the digits again, we finally arrive at the answer 9.

Let's try another number, 867. Doubling it gives 1,934, multiplying by 6 yields 10,404, adding the digits one last time we again arrive at the answer 9. Is this just a coincidence?

Ask your students to compose a list of three-digit numbers using this process, thereby giving them good practice in addition and single-digit multiplication and in following a formula. As you and they might expect, they will always end up with the number 9.

## 55. CALCULATION OF Π AND FRACTIONS (H, J)

Near the end of the sixth grade, students usually come upon the mystery of irrational numbers when they are taught something about finding the circumference and area of a circle.

They are told that the number π is used to find the circumference of a circle by the formula $C = 2 \pi r$ where r is the radius of the circle, and that this number is also used to find the area of the circle by the formula $A = \pi r^2$.

The number n is defined to be approximately equal to 22/7 or 3.14, and is further explained to be an irrational number, meaning that its decimal expansion is interminable and non-repeating.

The first 10 decimal places of π are 3.1415926535...

However, we can get some wonderful approximations to π by adding, subtracting and multiplying sequences of fractions.

**Stage 1**

Have your students apply the formula n = 4 x (1 − 1/3 + 1/5 − 1/7 + 1/9 − 1/11 + …).

Let's try to find the first few terms: 1 − 1/3 = 2/3 and 4 x 2/3 = 8/3 = 2.666..., not a very good approximation.

Continuing, 1 − 1/3 + 1/5 = 2/3 + 1/5 = 10/15 + 3/15 = 13/15 and 4 x 13/15 = 3.466..., a better approximation.

Going a little further, 1 − 1/3 + 1/5 − 1/7 + 1/9 = 13/15 − 1/7 + 1/9 = 273/315 + 45/315 + 35/315 = 263/315 and 4 x 263/315 = 1052/315 = 3.33666..., which is a still better approximation.

Have your daring students continue this approximation until they are able to get to the first digit after the decimal point correctly.

**Stage 2**

Another fractions formula for π is $\pi^2$ = 6 x ($1^2$ + $(1/2)^2$ + $(1/3)^2$ + $(1/4)^2$ + ...) or 6 x (1 + (1/4) + (1/9) + (1/16) + …).

We can estimate $\pi^2$ to four decimal places as 3.14 x 3.14 = 9.8696. Again, trying to find a few terms, we get: 1 + 1/4 = 5/4 and 5/4 x 6 = 15/2 = 7.5, not a very good approximation.

Continuing, $1^2$ + $(1/2)^2$ + $(1/3)^2$ = 5/4 + 1/9 = 45/36 + 4/36 = 49/36 and 49/36 x 6 = 8.166..., a somewhat better approximation.

One more time, 1 + $(1/2)^2$ + $(1/3)^2$ + $(1/4)^2$ = 49/36 + 1/16 = 196/144 + 9/144 = 205/144 and 205/144 x 6 = 8.541....

It will take a while for these approximations to get close to $\pi^2$. Challenge your students with just getting the number part 9 for $\pi^2$, which is equivalent to getting the whole number part 3 for π.

These fractional estimations for n are not only truly excellent practice in fraction skills, but lay a groundwork for the sophisticated mathematics of infinite series—which the students may see later if they eventually study Calculus.

## Stage 3

A final problem for interested students to work out is the formula for the sum of the powers of a fraction. For example, let's start with the following: $1 + 1/2 + (1/2)^2 + (1/2)^3 + (1/2)^4 + (1/2)^5 = 1/2 + 1/4 + 1/8 + 1/16 + 1/32$.

This is equivalent to $1 + 8/16 + 4/16 + 2/16 + 1/16 = 1 + 15/16$. If we take another term, we would have $1 + 16/32 + 8/32 + 4/32 + 2/32 + 1/32 = 1 + 31/32$. Continuing this, another step would leave us with $1 + 63/64$.

It should be obvious that we seem to be getting closer and closer to 2, and it is a fact that we can write 2 as $1/_{1-1/2} = 1/_{1/2}$.

The general formula states that if x is less than 1, $1 + x + x^2 + x^3 + ... = 1/_{1-x}$. Have your students try this out for a few different fractions, such as 1/3, 1/4, 2/3, and 3/4.

Let's see how 2/3 would work: $1 + 2/3 + 4/9 + 8/27 + 16/81 + 32/243 = 1 + 162/243 + 108/243 + 72/243 + 48/243 + 32/243 = 665/243 = 2 + 179/243$ and $1/_{1-2/3} = 1/_{1/3} = 3$. It will take quite a while to get very close to 3, but it will eventually happen—which is the meaning behind the notion of a mathematical limit.

With this problem we conclude Chapter 1 of this book, *Recreational Number Theory Problems*, and move on to the ideas for games in Chapters 2, 3, and 4, to practice and enhance the skills learned through working on these problems.

# Games of Recreational Number Theory: Skill Levels Through Addition, Subtraction, and Multiplication

**Abstract:** Chapter 2 consists of the games that require the skill levels through addition, subtraction, and multiplication. These games consist of The Tri Game, The Fib-Tri Game, The Multiplicative Persistence Game, The Tri-Square Game, The Tri-Squar- -Cube Game, and The Kaprekar Number Game. These games are in general appropriate for children in grades 1, 2, and 3, depending on the children's arithmetic skills level. However, they can be used as enjoyable recreational games for children of all ages, inclusive of mathematically gifted children. See the sections "Introduction to the Games" and "Game Ideas from Teachers at Numberama Teacher Workshops" in the beginning of this book, and **some of the teacher and teacher education Numberama workshop participant responses in the Appendix** for relevant information about playing these games effectively with children, as well as how these games were developed.

**Keywords:** Recreational number theory games, The Fib-Tri game, The kaprekar number game, The multiplicative persistence game, The Tri game, The Tri-square game, The Tri-square-cube game.

Fig. (9). The tri game (b).

## RULES FOR THE TRI GAME (FIG. 9)

1. Each player tosses die to determine order of turns and order of choice of banker.
2. Player tosses die and lands on number space. He/she must say the appropriate Triangular number. The Triangular numbers are the numbers in the sequence 1, 3, 6, 10, 15, 21, 28, 36, ... obtained by adding one more than the difference of the previous two numbers to the last number. The next number in the sequence would be 45 since $36 - 28 = 8$, $8 + 1 = 9$, and $36 + 9 = 45$. Thus if a player lands on 9, he/she must say the 9th Triangular number, which is 45.
3. The player whose turn is next looks at the back of the number card to see if the original player is correct. If he/she is correct, the original player gets that amount of money, *i.e.*, the Triangular number, from the bank.
4. If a player lands on GO exactly, he/she receives $100 and the game ends. If a player passes GO, he/she goes through the procedure for the number space he/she landed on; then the game ends. The player with the most money is the winner.

### Number Cards for the Tri Game

| | |
|:---:|:---:|
| 1 | 1 |
| 2 | 3 |
| 3 | 6 |
| 4 | 10 |
| 5 | 15 |
| 6 | 21 |
| 7 | 28 |
| 8 | 36 |
| 9 | 45 |
| 10 | 55 |
| 11 | 66 |
| 12 | 78 |
| 13 | 91 |
| 14 | 105 |
| 15 | 120 |
| 16 | 136 |
| 17 | 153 |
| 18 | 171 |
| 19 | 190 |

*Contd.....*

| 20 | 210 |
|----|-----|
| 21 | 231 |
| 22 | 253 |
| 23 | 276 |
| 24 | 300 |
| 25 | 325 |
| 26 | 351 |
| 27 | 378 |

**Fig. (10).** The fib-tri game (b).

## RULES FOR THE FIB-TRI GAME (FIG.10)

1. Each player tosses die to determine order of turns and order of choice of banker.

2. The player tosses die and lands on number space. He/She must say the appropriate Triangular number and Fibonacci number. The Triangular numbers are the numbers in the sequence 1, 3, 6, 10, 15, 21, 28, 35, ... obtained by adding one more than the difference of the previous two numbers to the last number. The next Triangular number in the sequence would be 45, since 36 − 28 = 8, 8 + 1 = 9, and 36 + 9 = 45. Thus, if a player lands on 9, he/she must say the 9th Triangular number is 45. The Fibonacci numbers are the numbers in the sequence 1,1, 3, 5, 8, 13, 21, 34, ... obtained by adding the sum of the previous

two numbers to obtain the next number. The next Fibonacci number is 55, since 21 + 34 = 55. Thus, the tenth Fibonacci number is 55.

3. The player whose turn is next looks at the back of the number card to see if original player is correct for both the Triangular number and the Fibonacci number. If both numbers are correct, player gets amount of money circled, (*i.e.*, the Triangular number) from the bank.

4. If a player lands on GO exactly, he/she receives $100 and the game ends. If a player passes GO, he/she goes through the procedure for the number space he/she landed on: then the game ends. The player with the most money is the winner.

### Number Cards for the Fib-Tri Game

| | | |
|---|---|---|
| 1 | T1 | F1 |
| 2 | T3 | F1 |
| 3 | T6 | F2 |
| 4 | T10 | F3 |
| 5 | T15 | F5 |
| 6 | T21 | F8 |
| 7 | T28 | F13 |
| 8 | T36 | F21 |
| 9 | T45 | F34 |
| 10 | T55 | F55 |
| 11 | T66 | F89 |
| 12 | T78 | F144 |
| 13 | T91 | F233 |
| 14 | T105 | F377 |
| 15 | T120 | F610 |
| 16 | T136 | F987 |
| 17 | T153 | F1,597 |
| 18 | T171 | F2,584 |
| 19 | T190 | F4,181 |
| 20 | T210 | F6,765 |
| 21 | T231 | F10,946 |
| 22 | T253 | F17,711 |
| 23 | T276 | F28,657 |
| 24 | T300 | F46,368 |
| 25 | T325 | F75,025 |

*Contd.....*

| 26 | T351 | F121,393 |
| 27 | T378 | F196,418 |

| 97 | 85 | 98 | 91 | 68 | 86 | 93 | 72 |
|----|----|----|----|----|----|----|----|
| 88 | | | | | | | 75 |
| 77 | | | | | | | 87 |
| 99 | | THE | MULTIPLICATIVE | PERSISTENCE | GAME | | 59 |
| 96 | | | | | | | 73 |
| 76 | | | | | | | 66 |
| 89 | | | | | | | 53 |
| GO | 17 | 25 | 37 | 42 | 19 | 39 | 45 |

**Fig. (11).** The multiplicative persistence game (b).

## RULES FOR THE MULTIPLICATIVE PERSISTENCE GAME (FIG. 11)

1. Each player tosses die to determine order of turns and order of choice of banker.
2. Player tosses die and lands on number space. He/she must multiply the digits of the number space together, then do it again, continuing until a single-digit number is reached. The number of steps it takes to get a single-digit number is the "multiplicative persistence" of the original number. For example: the sequence for 78 is 78 – 56 – 30 – 0, so 78 has multiplicative persistence of 3.
3. The player whose turn is next looks at the back of the number card. If the original player has correctly said both the multiplicative persistence and all the numbers in the sequence, he/she then must multiply the original number by the multiplicative persistence. For example: 78 x 3 = 234.
4. If the player's answer is correct, he/she gets that amount of money from the bank. In the case of a number with multiplicative persistence of 4, $100 bonus is added to the amount of money.
5. If a player lands on GO exactly, he/she gets $100 and the game ends; the player with the most money is the winner. If a player passes GO, he/she does the usual procedure for the number space landed on, and then the game ends; again the

player with the most money is the winner.

## Number Cards for the Multiplicative Persistence Game

| 17 | 17-7 | 1 | $17 |
|---|---|---|---|
| 25 | 25-10-0 | 2 | $50 |
| 37 | 37-21-1 | 2 | $74 |
| 42 | 42-8 | 1 | $42 |
| 19 | 19-9 | 1 | $19 |
| 39 | 39-27-14-4 | 3 | $117 |
| 45 | 45-20-0 | 2 | $90 |
| 53 | 53-15-5 | 2 | $106 |
| 66 | 66-36-18-8 | 3 | $198 |
| 73 | 73-21-2 | 2 | $146 |
| 59 | 59-45-20-0 | 3 | $177 |
| 87 | 87-56-30-0 | 3 | $261 |
| 75 | 75-35-15-5 | 3 | $225 |
| 72 | 72-14-4 | 2 | $144 |
| 93 | 93-27-14-4 | 3 | $279 |
| 86 | 86-48-32-6 | 3 | $258 |
| 68 | 68-48-32-6 | 3 | $204 |
| 91 | 91-9 | 1 | $91 |
| 98 | 98-72-14-4 | 3 | $294 |
| 85 | 85-40-0 | 2 | $170 |
| 97 | 97-63-18-8 | 3 | $291 |
| 88 | 88-64-24-8 | 3 | $264 |
| 77 | 77-49-36-18-8 | 4 | $408 |
| 99 | 99-81-8 | 2 | $198 |
| 96 | 96-54-20-0 | 3 | $288 |
| 76 | 76-42-8 | 2 | $152 |
| 89 | 89-72-14-4 | 3 | $267 |

| 22 | 21 | 20 | 19 | 18 | 17 | 16 | 15 |
|----|----|----|----|----|----|----|----|
| 23 |    |    |    |    |    |    | 14 |
| 24 |    |    |    |    |    |    | 13 |
| 25 |    | THE |    |    |    |    | 12 |
| 26 |    | TRI-SQUARE |    |    |    |    | 11 |
| 27 |    | GAME |    |    |    |    | 10 |
| 28 |    |    |    |    |    |    | 9 |
| GO | 2 | 3 | 4 | 5 | 6 | 7 | 8 |

**Fig. (12).** The tri-square game (d).

## RULES FOR THE TRI-SQUARE GAME (FIG. 12)

1. Each player rolls die to determine order of turns and order of choice of banker.
2. Player rolls die and lands on number space. He/she must square the number he/she lands on, *i.e.* multiply the number by itself, and then find two Triangular numbers that add up to the square of the number. The Triangular numbers are the numbers in the sequence 1, 3, 6, 10, 15, 21, ... Each number in the sequence is determined by adding 1 more than the difference of the previous two numbers to the last number.
3. The player whose turn is next looks at the back of the number card to see if the original player is correct. As an example, $6^2 = 36$ and $15 + 21 = 36$. If player is correct, he/she gets the amount of money equal to the larger of the two Triangular numbers. In the above example, he/she would get $21.
4. If player lands on GO exactly, he/she receives $100, has the chance to do number space #28, and then the game ends. If player passes GO, he/she goes through the procedure for the number space he/she landed on; then the game ends. The player with the most money is the winner.

### Number Cards for the Tri-Square Game

| 2 | $2^2 = 4$ | $1+3 = 4$ | $3 |
|---|-----------|-----------|-----|
| 3 | $3^2 = 9$ | $3+6 = 9$ | $6 |
| 4 | $42= 16$ | $6+10 = 16$ | $10 |

*Contd.....*

| | | | |
|---|---|---|---|
| 5 | $5^2 = 25$ | $10+15 = 25$ | $15 |
| 6 | $6^2 = 36$ | $15+21 = 36$ | $21 |
| 7 | $7^2 = 49$ | $21+28 = 49$ | $28 |
| 8 | $8^2 = 64$ | $28+36 = 64$ | $36 |
| 9 | $9^2 = 81$ | $36+45 = 81$ | $45 |
| 10 | $10^2 = 100$ | $45+55 = 100$ | $55 |
| 11 | $11^2 = 121$ | $55+66 = 121$ | $66 |
| 12 | $12^2 = 144$ | $66+78 = 144$ | $78 |
| 13 | $13^2 = 169$ | $78+91 = 169$ | $91 |
| 14 | $14^2 = 196$ | $91+105 = 196$ | $105 |
| 15 | $15^2 = 225$ | $105+120 = 225$ | $120 |
| 16 | $16^2 = 256$ | $120+136 = 256$ | $136 |
| 17 | $17^2 = 289$ | $136+153 = 289$ | $153 |
| 18 | $18^2 = 324$ | $153+171 = 324$ | $171 |
| 19 | $19^2 = 361$ | $171+190 = 361$ | $190 |
| 20 | $20^2 = 400$ | $190+210 = 400$ | $210 |
| 21 | $21^2 = 441$ | $210+231 = 441$ | $231 |
| 22 | $22^2 = 484$ | $231+253 = 484$ | $253 |
| 23 | $23^2 = 529$ | $253+276 = 529$ | $276 |
| 24 | $24^2 = 576$ | $276+300 = 576$ | $300 |
| 25 | $25^2 = 625$ | $300+325 = 625$ | $325 |
| 26 | $26^2 = 676$ | $325+351 = 676$ | $351 |
| 27 | $27^2 = 729$ | $351+378 = 729$ | $378 |
| 28 | $28^2 = 784$ | $378+406 = 784$ | $406 |

**Fig. (13).** The tri-square-cube game (d).

## RULES FOR THE TRI-SQUARE-CUBE GAME (FIG.13)

1. Each player rolls die to determine order of turns and order of choice of banker.
2. Player rolls die and lands on number space. He/she must say each number cubed, from 1 until the number landed on, and add up all of these cubes. Then he/she must say the corresponding Triangular number, square it, and show that it is the same number as the previous sum of cubes. For example, if player lands on 5, $1^3 + 2^3 + 3^3 + 4^3 + 5^3 = 1 + 8 + 27 + 64 + 125 = 225$ and the fifth Triangular number is 15 (1, 3, 6, 10, 15) and $15^2 = 225$.
3. The player whose turn is next looks at back of number card to see if original player is correct. If all answers are correct, player gets the number circled, which is the Triangular number, of dollars from the bank.
4. If player lands on GO exactly, he/she receives $100 and the game ends. If player passes GO, he/she goes through the procedure for the number space he/she landed on; then the game ends. The player with the most money is the winner.

### Number Cards for the Tri-Square-Cube Game

| | | | |
|---|---|---|---|
| 1 | $1^2 = 1$ | $1^3 = 1$ | $1 |
| 2 | $3^2 = 9$ | $1^3 + 2^3 = 1 + 8 = 9$ | $3 |
| 3 | $6^2 = 36$ | $1^3 + 2^3 + 3^3 = 1 + 8 + 27 = 36$ | $6 |
| 4 | $10^2 = 100$ | $1^3 + 2^3 + 3^3 + 4^3 = 1 + 8 + 27 + 64 = 100$ | $10 |
| 5 | $15^2 = 225$ | $1^3 + 2^3 + 3^3 + 4^3 + 5^3 = 1 + 8 + 27 + 64 + 125 = 225$ | $15 |
| 6 | $21^2 = 441$ | $1^3 + 2^3 + 3^3 + 4^3 + 5^3 + 6^3 = 1 + 8 + 27 + 64 + 125 + 216 = 441$ | $21 |
| 7 | $28^2 = 784$ | $1^3 + 2^3 + 3^3 + 4^3 + 5^3 + 6^3 + 7^3 = 1 + 8 + 27 + 64 + 125 + 216 + 343 = 748$ | $28 |
| 8 | $36^2 = 1,296$ | $1^3 + 2^3 + 3^3 + \dots 8^3 = 1 + 8 + 27 + 64 + 125 + 216 + 343 + 512 = 1,296$ | $36 |

*Contd.....*

| 9 | 452 = 2,025 | $1^3+2^3+3^3+...9^3 = 1+8+27+64+125+216+343+512+729 = 2025$ | $45 |
|---|---|---|---|
| 10 | 552 = 3,025 | $1^3+2^3+3^3+...10^3 = 1+8+27+64+125+216+343+512+729+1000 = 3025$ | $55 |
| 11 | 662 = 4,356 | $1^3+2^3+3^3+...11^3 = 1+8+27+64+125+216+343+512+729+1000+1331 = 4356$ | $66 |
| 12 | 782 = 6,084 | $1^3+2^3+3^3+...12^3 = 1+8+27+64+125+216+343+512+729+1000+1331+1728 = 6084$ | $78 |
| 13 | 912 = 8,281 | $1^3+2^3+3^3+...13^3 = 1+8+27+64+125+216+343+512+729+1000+1331+1728+2197 = 8281$ | $91 |
| 14 | 1052 = 11,025 | $1+2^3+3^3+...14^3 = 1+8+27+64+125+216+343+512+729+1000+1331+1728+2197+2744 = 11025$ | $105 |
| 15 | 1202 = 14,400 | $1^3+2^3+3^3+...5^3 = 1+8+27+64+125+216+343+512+729+1000+1331+1728+2197+2744+3375 = 14400$ | $120 |
| 16 | 1362 = 18,496 | $1^3+2^3+3^3+...16^3 = 1+8+27+64+125+216+343+512+729+1000+1331+1728+2197+2744+3375+4096 = 18496$ | $136 |
| 17 | 1532 = 23,409 | $1^3+2^3+3^3+...17^3 = 1+8+27+64+125+216+343+512+729+1000+1331+1728+2197+2744+3375+4096+4913 = 23409$ | $153 |
| 18 | 1712 = 29,241 | $1^3+2^3+3^3+...18^3 = 1+8+27+64+125+216+343+512+729+1000+1331+1728+2197+2744+3375+4096+4913+5832 = 29241$ | $ 171 |
| 19 | 1902 = 36,100 | $1^3+2^3+3^3+...19^3 = 1+8+27+64+125+216+343+512+729+1000+1331+1728+2197+2744+3375+4096+4913+5832+6859 = 36100$ | $ 190 |
| 20 | 2102 = 44,100 | $1^3+2^3+3^3+...20^3 = 1+8+27+64+125+216+343+512+729+1000+1331+1728+2197+2744+3375+4096+4913+5832+6859+8000 = 44100$ | $210 |
| 21 | 2312 = 53,361 | $1^3+2^3+3^3+...21^3 = 1+8+27+64+125+216+343+512+729+1000+1331+1728+2197+2744+3375+4096+4913+5832+6859+8000+9261 = 53361$ | $231 |
| 22 | 2532 = 64,009 | $1^3+2^3+3^3+...22^3 = 1+8+27+64+125+216+343+512+729+1000+1331+1728+2197+2744+3375+4096+4913+5832+6859+8000+9261+10648 = 64009$ | $253 |
| 23 | 2762 = 76,176 | $1^3+2^3+3^3+...23^3 = 1+8+27+64+125+216+343+512+729+1000+1331+1728+2197+2744+3375+4096+4913+5832+6859+8000+9261+10648+12167 = 76176$ | $276 |
| 24 | 3002 = 90,000 | $1^3+2^3+3^3+...24^3 = 1+8+27+64+125+216+343+512+729+1000+1331+1728+2197+2744+3375+4096+4913+5832+6859+8000+9261+10648+12167+13824 = 90000$ | $300 |
| 25 | 3252 = 105,625 | $1^3+2^3+3^3+...25^3 =$ $1+8+27+64+125+216+343+512+729+1000+1331+1728+2197+2744+3375+4096+4913+5832+6859+8000+9261+10648+12167+13824+15625 = 105625$ | $325 |
| 26 | 3512 = 123,201 | $1^3+2^3+3^3+...26^3 =$ $1+8+27+64+125+216+343+512+729+1000+1331+1728+2197+2744+3375+4096+4913+5832+6859+8000+9261+10648+12167+13824+15625+17576 = 123201$ | $351 |
| 27 | 3782 = 142,884 | $1^3+2^3+3^3+...27^3 =$ $1+8+27+64+125+216+343+512+729+1000+1331+1728+2197+2744+3375+4096+4913+5832+6859+8000+9261+10648+12167+13824+15625+17576+19683 = 142884$ | $378 |

| 87 | 84 | 81 | 79 | 76 | 73 | 71 | 70 |
|----|----|----|----|----|----|----|----|
| 89 | | | | | | | 69 |
| 91 | | | | | | | 66 |
| 94 | | THE KAPREKAR NUMBER GAME | | | | | 64 |
| 96 | | | | | | | 61 |
| 97 | | | | | | | 55 |
| 99 | | | | | | | 54 |
| GO | 35 | 41 | 44 | 45 | 50 | 52 | 53 |

**Fig. (14).** The kaprekar number game (d).

# RULES FOR THE KAPREKAR NUMBER GAME (FIG.14)

1. Each player rolls die to determine order of turns and order of choice of banker.
2. Player rolls die and lands on number space. He/she must square the number landed on, *i.e.*, multiply it by itself, then add each pair of digits together from the resulting four-digit number. For example $35^2 = 1,225$ and $12 + 25 = 37$. If the sum of the 2 pairs of digits equals the original number, then the number is called a Kaprekar number. Thus 35 is not a Kaprekar number since $35^2 = 1,225$ and $12 + 25 = 37$ which is not equal to 35.
3. The player whose turn is next looks at back of number card. If the answer is correct and the number is not a Kaprekar number, the player receives the sum of the pairs of digits amount of money from the bank. If the number is a Kaprekar number and the player answers correctly, stating it as a Kaprekar number, he/she receives sum of pairs of digits plus an additional $200 amount of money from the bank. If player does not say it is a Kaprekar number, he/she receives only sum of pairs of digits amount of money from the bank.
4. If player lands on GO exactly, he/she receives $100 and the game ends. If player passes GO, he/she goes through the procedure for the number space he/she landed on; then the game ends. The player with the most money is the winner.

## Number Cards for the Kaprekar Number Game

| | | | |
|---|---|---|---|
| 35 | $35^2 = 1225$ | | $37 |
| 41 | $41^2 = 1681$ | | $97 |
| 44 | $44^2 = 1936$ | | $55 |
| 45 | $45^2 = 2025$ | Kaprekar Number | $245 |
| 50 | $50^2 = 2500$ | | $25 |
| 52 | $52^2 = 2704$ | | $31 |
| 53 | $53^2 = 3809$ | | $37 |
| 54 | $54^2 = 2916$ | | $45 |
| 55 | $55^2 = 3025$ | Kaprekar Number | $255 |
| 61 | $61^2 = 3721$ | | $58 |
| 64 | $64^2 = 4096$ | | $136 |
| 66 | $66^2 = 4356$ | | $99 |
| 69 | $69^2 = 4761$ | | $108 |
| 70 | $70^2 = 4900$ | | $49 |
| 71 | $71^2 = 5041$ | | $91 |
| 73 | $73^2 = 5329$ | | $82 |
| 76 | $76^2 = 5776$ | | $133 |

*Contd.....*

| 79 | $79^2 = 6241$ | | $103 |
|---|---|---|---|
| 81 | $81^2 = 6561$ | | $126 |
| 84 | $84^2 = 7056$ | | $126 |
| 87 | $87^2 = 7569$ | | $144 |
| 89 | $89^2 = 7921$ | | $100 |
| 91 | $91^2 = 8281$ | | $163 |
| 94 | $94^2 = 8836$ | | $124 |
| 96 | $96^2 = 9216$ | | $108 |
| 97 | $97^2 = 9409$ | | $103 |
| 99 | $99^2 = 9801$ | Kaprekar Numbe | $299 |

# Games of Recreational Number Theory: Skill Levels Through Multiplication and Division

**Abstract:** Chapter 3 consists of the games that require the skill levels through multiplication and division. These games consist of The Prime Number Game, The Perfect Number Game, The Semi-Perfect Number Game, The Powerful Number Game, The Divisor Game, The Sum of Squares Game, The Syracuse Algorithm Game, and The Fib-Tri-Prime Game. The games in Chapter 3 are in general appropriate for children in grades 4, 5, and 6, depending on their skills levels, inclusive of gifted children. The Perfect Number Game has been especially popular with children, in particular because of the stimulating open problems about perfect numbers in mathematics; *i.e.* are there infinitely many perfect numbers and does there exist an odd perfect number (see the Perfect Number problem section in Chapter 1). The games in Chapter 3 are excellent teaching devices to motivate children to practice their multiplication and division skills in an enjoyable way, in addition to enhancing their creative thinking capacities (see some of the teacher and teacher education Numberama workshop participant responses in Appendix).

**Keywords:** The divisor game, The Fib-Tri-Prime game, The perfect number game, The powerful number game, The prime number game, The semi-perfect number game, The sum of squares game, The syracuse algorithm game.

| 21 | 20 | 19 | 18 | 17 | 16 | 15 | 14 |
|----|----|----|----|----|----|----|----|
| 22 |    |    |    |    |    |    | 13 |
| 23 |    |    |    |    |    |    | 12 |
| 24 |    | THE |  |    |    |    | 11 |
| 25 |    | PRIME NUMBER |  |  |    |    | 10 |
| 26 |    | GAME |  |    |    |    | 9 |
| 27 |    |    |    |    |    |    | 8 |
| GO | 1 | 2 | 3 | 4 | 5 | 6 | 7 |

Fig. (15). The prime number game (g).

## RULES FOR THE PRIME NUMBER GAME

1. Each player rolls die to determine order of turns and order of choice of banker.
2. Player rolls die and lands on number space. He/she must say the corresponding prime number (P), twin-prime number (T), and semi-prime number (S) (see the Definitions below). For example, if a player lands on 5, he/she must say the fifth prime number, the fifth twin-prime number, and the fifth semi-prime number.
3. The player whose turn is next looks at the back of the appropriate number card to see if the original player is correct. If the player is correct, he/she must multiply the sum of all the digits in the three numbers he/she landed on by the number space. If he/she is again correct, he/she gets that amount of money from the bank.
4. If player lands on GO exactly, he/she receives $100 and the game ends. If player passes GO, he/she goes through the procedure for the number space he/she landed on; then the game ends. The player with the most money is the winner.

## DEFINITIONS FOR THE PRIME NUMBER GAME (FIG.15)

*Prime Number:* a number that is only divisible by the number itself and 1. Examples of prime numbers are 3, 7, 11, 17, 19, 29.

*Twin-Prime Number,* prime numbers that differ from another prime number by either 1 or 2. Except for 2 and 3, all twin-prime numbers differ by 2 from another prime number. Examples of twin-prime numbers are 3, 5, 7, 11, 13, 17, 19, 39, 31, 41, 43.

*Semi-Prime Number,* numbers that have exactly 4 divisors, including the number itself and 1. Examples of semi-prime numbers are 8, 15, 21, 33, 35, 62.

### Number Cards for the Prime Number Game

| 1 | P2 | T2 | S6 | $3 |
|---|-----|-----|-----|-----|
| 2 | P3 | T3 | S8 | $6 |
| 3 | P5 | T5 | S10 | $12 |
| 4 | P7 | T7 | S14 | $16 |
| 5 | P11 | T11 | S15 | $30 |
| 6 | P13 | T13 | S21 | $36 |
| 7 | P17 | T17 | S22 | $42 |
| 8 | P19 | T19 | S26 | $48 |
| 9 | P23 | T29 | S27 | $54 |

*Contd.....*

| 10 | P29 | T31 | S33 | $60 |
|----|-----|-----|-----|-----|
| 11 | P31 | T41 | S34 | $66 |
| 12 | P37 | T43 | S35 | $72 |
| 13 | P41 | T59 | S38 | $78 |
| 14 | P43 | T61 | S39 | $84 |
| 15 | P47 | T71 | S46 | $90 |
| 16 | P53 | T73 | S51 | $96 |
| 17 | P59 | T101 | S55 | $119 |
| 18 | P61 | T103 | S57 | $126 |
| 19 | P67 | T107 | S65 | $133 |
| 20 | P71 | T109 | S69 | $140 |
| 21 | P73 | T137 | S74 | $147 |
| 22 | P79 | T139 | S77 | $154 |
| 23 | P83 | T149 | S82 | $161 |
| 24 | P89 | T151 | S85 | $168 |
| 25 | P97 | T171 | S86 | $175 |
| 26 | P101 | T181 | S87 | $208 |
| 27 | P103 | T191 | S91 | $216 |

**Fig. (16).** The perfect number game (g).

## RULES FOR THE PERFECT NUMBER GAME (FIG. 16)

1. Each player rolls die to determine order of turns and order of choice of banker.
2. Player rolls die and lands on number space. He/she must write down correctly all the proper divisors of the number (*i.e.*, all divisors of the number other than the number itself) and say if the number is abundant, deficient, or perfect. The definitions are as follows: a number is abundant if the sum of its proper divisors is greater than the number, deficient if the sum of its proper divisors is less than the number, and perfect if the sum of its proper divisors equals the number. For example, 40 is abundant because its proper divisors are 1, 2, 20, 4, 10, 5, and 8, and their sum is 50; 35 is deficient because its proper divisors are 1, 5, and 7, and their sum is 13.
3. The player whose turn is next looks at back of number card to see if the first player's answers are correct. If all answers are correct, the player gets money from the bank based upon the following formula:

|  | Yellow | Blue | Green | Red |
|---|---|---|---|---|
| Deficient | $5 | $10 | $15 | $20 |
| Abundant | $10 | $20 | $30 | $40 |
| Perfect | $50 | $100 | $150 | $200 |

There are 12 abundant numbers, 12 deficient numbers, and 3 perfect numbers on the gameboard.

4. If player lands on GO exactly, he/she receives $50 and the game ends. If player passes GO, he/she goes through the procedure for the number space he/she landed on; then the game ends. The player with the most money is the winner.

### Number Cards for the Perfect Number Game

| | | |
|---|---|---|
| 5 | Deficient | 1 = 1 |
| 6 | Perfect | 1+2+3 = 6 |
| 8 | Deficient | 1+2+4 = 7 |
| 12 | Abundant | 1+2+3+4+6 = 16 |
| 15 | Deficient | 1+3+5 = 9 |
| 18 | Abundant | 1+2+3+6+9 = 21 |
| 20 | Abundant | 1+2+4+5+10 = 22 |
| 22 | Deficient | 1+2+11 = 14 |
| 24 | Abundant | 1+2+3+4+6+8+12 = 36 |
| 26 | Deficient | 1+2+13 = 16 |

*Contd.....*

| 28 | Perfect | 1+2+4+7+14 = 28 |
|---|---|---|
| 30 | Abundant | 1+2+3+5+6+10+15 = 42 |
| 32 | Deficient | 1+2+4+8+16 = 31 |
| 50 | Deficient | 1+2+5+10+25 = 43 |
| 100 | Abundant | 1+2+4+5+10+20+25+50 = 117 |
| 128 | Deficient | 1+2+4+8+16+32+64 = 127 |
| 150 | Abundant | 1+2+3+5+6+10+75+50+30+25+15 = 222 |
| 170 | Deficient | 1+85+2+5+34+10+17 = 154 |
| 196 | Abundant | 1+2+98+4+49+14+7+28 = 203 |
| 200 | Abundant | 1+2+4+5+10+20+25+50+100+40+8 = 265 |
| 250 | Deficient | 1+2+125+5+50+10+25 = 218 |
| 275 | Deficient | 1+5+55+11+25 = 97 |
| 300 | Abundant | 1+2+150+3+100+5+60+4+75+6+50+10+30+12+25+15+20 = 568 |
| 380 | Abundant | 1+190+2+4+95+10+38+20+19+5+76 = 460 |
| 442 | Deficient | 1+2+221+13+34+26+17 = 314 |
| 496 | Perfect | 1+2+248+4+124+8+62+16+31 = 496 |
| 500 | Abundant | 1+2+250+5+100+10+50+20+25+4+125 = 592 |

**Fig. (17).** The semi-perfect number game (g).

## RULES FOR THE SEMI-PERFECT NUMBER GAME (FIG. 17)

1. Each player rolls die to determine order of turns and order of choice of banker.
2. Player rolls die and lands on number space. Player must state if number is abundant—which means the sum of its proper divisors is greater than the number, or deficient—which means the sum of its proper divisors is less than the number, and must also say all of the number's proper divisors (a proper divisor is a divisor other than the number itself). If the number is abundant, the player must determine a combination of some of the divisors that add up to the number; the number is thereby defined to be a semi-perfect number. If there are no combinations of divisors that add up to an abundant number, the number is called "weird," meaning it is abundant but not semi-perfect.
3. The player whose turn is next looks at back of number card to see if original player is correct. A player only needs to say one combination of divisors that add up to the number if it is semi-perfect. For example, 12 is abundant since the proper divisors of 12 are 1, 2, 3, 4, and 6, adding up to 16. The combination of divisors 6, 4, and 2 add up to 12; the combination of divisors 6, 3, 2, and 1 also add up to 12. If all answers are correct, the player gets money from the bank based upon the following formula:

|              | Green | Yellow | Red   | Blue  |
|--------------|-------|--------|-------|-------|
| Deficient    | $5    | $10    | $15   | $20   |
| Semi-Perfect | $20   | $40    | $60   | $80   |
| Weird        | $50   | $100   | $150  | $200  |

It should be noted that there is only one "weird" two-digit number, and it is on the gameboard.

4.  If player lands on GO exactly, he/she receives $50 and the game ends. If player passes GO, he/she goes through the procedure for the number space he/she landed on; then the game ends. The player with the most money is the winner.

### Number Cards for the Semi-Perfect Number Game

| 4  | Deficient    | 1+2 = 3 |
|----|--------------|---------|
| 10 | Deficient    | 1+2+5 = 8 |
| 12 | Semi-Perfect | 1+2+3+4+6 = 12, 2+4+6 = 12 |
| 16 | Deficient    | 1+2+4+8 = 15 |
| 20 | Semi-Perfect | 1+2+4+5+10 = 22, 10+5+4+1 = 20 |
| 24 | Semi-Perfect | 1+2+3+4+6+8+12 = 36, 12+8+4 = 24, 12+6+4+2 = 24, 8+6+4+3+2+1= 24 |
| 30 | Semi-Perfect | 1+2+3+5+6+10+15 = 42, 15+10+5 = 30, 15+10+3+2 = 30 |

*Contd.....*

| 36 | Semi-Perfect | 1+2+3+4+6+9+12+18 = 55, 18+12+6 = 36, 18+9+6+3 = 36, 12+9+6+4+3+2 = 36 |
|---|---|---|
| 40 | Semi-Perfect | 1+2+4+5+8+10+20 = 50, 20+10+5+4+1 = 40 |
| 44 | Deficient | 1+2+4+11+22 = 40 |
| 48 | Semi-Perfect | 1+2+3+4+6+8+12+16+24 = 76, 24+16+6+2 = 48, 24+16+6+2 = 48 |
| 50 | Deficient | 1+2+5+10+25 = 43 |
| 56 | Semi-Perfect | 1+2+4+7+8+14+28 = 64, 28+14+8+4+2 = 56 |
| 60 | Semi-Perfect | 1+2+3+4+5+6+10+12+15+20+30 = 108, 30+15+12+3 = 60, 30+12+10+5+3 = 60, 30+10+6+5+4+3+2 = 60 |
| 64 | Deficient | 1+2+4+8+16+32 = 63 |
| 68 | Deficient | 1+2+4+17+34 = 58 |
| 70 | Weird | 1+2+5+7+10+14+35 = 74 |
| 72 | Semi-Perfect | 1+2+3+4+6+8+12+18+24+36 = 114, 36+24+12 = 72, 24+18+12+9+8+1 = 72, 36+18+12+6 = 72 |
| 78 | Semi-Perfect | 1+2+3+6+13+26+39 = 90, 39+26+13 = 78 |
| 80 | Semi-Perfect | 1+2+4+5+8+10+16+20+40=106, 40+20+16+4 = 80, 40+20+10+8+2 = 80, 40+16+10+8+5+1 = 80 |
| 84 | Semi-Perfect | 1+2+3+4+6+7+12+21+28+42 = 143, 42+28+14 = 84, 42+21+14+7 = 84, 28+21+14+12+7+2 = 84 |
| 88 | Semi-Perfect | 1+2+4+8+11+22+44 = 92, 44+22+11+8+2+1 = 88 |
| 90 | Semi-Perfect | 1+2+3+6+9+10+15+30+45 = 121, 45+30+15 = 90, 45+30+10+3+2 = 90 |
| 92 | Deficient | 1+2+4+23+46 = 76 |
| 96 | Semi-Perfect | 1+2+3 4+6+8+12+16+24 32+48=156, 48+32+16 = 96, 48+16+12+8+6+2 = 96, 48+32+12+3+1 = 96 |
| 100 | Semi-Perfect | 1 +2+4+5+10+20+25+50 =117, 50+25+20+5 = 100, 50+25+20+4+1 = 100 |

| 300 | 288 | 275 | 255 | 250 | 225 | 200 | 195 |
|---|---|---|---|---|---|---|---|
| 324 | | | | | | | 180 |
| 350 | | | | | | | 144 |
| 400 | | | **THE** | | | | 128 |
| 441 | | | **POWERFUL** **NUMBER** **GAME** | | | | 99 |
| 483 | | | | | | | 88 |
| 511 | | | | | | | 63 |
| GO | 4 | 6 | 8 | 16 | 27 | 36 | 54 |

**Fig. (18).** The powerful number game (g).

## RULES FOR THE POWERFUL NUMBER GAME (FIG. 18)

1. The player rolls die to determine the order of turns and order of choice of banker.
2. The player rolls die and lands on number space. He/she must give the prime factorization of the number. For example, the prime factorization of $60 = 2 \times 30 = 2 \times 3 \times 10 = 2 \times 3 \times 2 \times 5 = 2^2 \times 3 \times 5$. If all primes in the number's prime factorization have exponents of 2 or greater, the number is called "powerful"; otherwise it is "not powerful." The number 60 is therefore not powerful, but $72 = 36 \times 2 = 6 \times 6 \times 2 = 3 \times 2 \times 3 \times 2 \times 2 = 2^3 \times 3^2$ is powerful. The player must then give the prime factorization of the next consecutive number. If the original number is $60 = 2^2 \times 3 \times 5$, he/she must say $61 = 61^1$, correctly saying whether each number is powerful or not powerful.
3. The player whose turn is next checks back of number card. If both prime factorizations are correct and labels of powerful and not powerful are correct, player receives money from the bank in the following way. If one number is powerful and the other number is not powerful, he/she receives the powerful number as money. If both numbers are not powerful, he/she receives half of the even "not powerful" number. If both numbers are powerful, he/she receives the sum of both numbers plus $200 bonus.
4. If player lands on GO exactly, he/she gets $100 and game ends; player with the most money is the winner. If player passes GO, he/she plays number space and then game ends.

### Number Cards for the Powerful Number Game

| | | | | |
|---|---|---|---|---|
| 4 | $4 = 2^2$P | $5 = 5^1$ N | | $4 |
| 6 | $6 = 2 \times 3$ N | $7 = 7^1$ N | | $3 |
| 8 | $8 = 2^3$P | $9 = 3^2$P | $8 + 9 + 200 =$ | $217 |
| 16 | $16 = 2^4$P | $17 = 17^1$ N | | $16 |
| 27 | $27 = 3^3$P | $28 = 2^2 \times 7$ N | | $27 |
| 36 | $36 = 2^2 \times 3^2$ P | $37 = 37^1$ N | | $36 |
| 54 | $54 = 3^3 \times 2$ N | $55 = 11 \times 5$ N | | $27 |
| 63 | $63 = 3^2 \times 7$ N | $64 = 2^6$ P | | $64 |
| 88 | $88 = 2^3 \times 11$ N | $89 = 89^1$ N | | $44 |
| 99 | $99 = 3^2 \times 11$ N | $100 = 5^2 \times 2^2$ P | | $100 |
| 128 | $128 = 2^7$ P | $129 = 3 \times 43$ N | | $128 |
| 144 | $144 = 3^2 \times 2^4$ P | $145 = 5 \times 29$ N | | $144 |
| 180 | $180 = 3^2 \times 2^2 \times 5^2$ N | $181 = 181^1$ N | | $90 |
| 195 | $195 = 13 \times 3 \times 5$ N | $196 = 7^2 \times 2^2$ P | | $196 |

*Contd*

| 200 | $200 = 2^3 \times 5^2$ P | $201 = 3 \times 67$ N | | $200 |
|-----|---------------------------|------------------------|------------------------|-------|
| 225 | $225 = 5^2 \times 3^2$ P | $226 = 113 \times 2$ N | | $225 |
| 250 | $250 = 5^3 \times 2$ N | $251 = 251$ N | | $125 |
| 255 | $255 = 5 \times 51$ N | $256 = 2^8$ P | | $256 |
| 275 | $275 = 5^2 \times 11$ N | $276 = 2^2 \times 3 \times 23$ N | | $138 |
| 288 | $288 = 2^5 \times 3^2$ P | $289 = 17^2$ P | $288 + 289 + 200 =$ | $777 |
| 300 | $300 = 5^2 \times 2^2 \times 3$ N | $301 = 7 \times 43$ N | | $150 |
| 324 | $324 = 3^4 \times 22$ P | $325 = 5^2 \times 13$ N | | $324 |
| 350 | $350 = 5^2 \times 2 \times 7$ N | $351 = 3^2 \times 3 \times 13$ N | | $175 |
| 400 | $400 = 5^2 \times 2^3$ P | $401 = 401^1$ N | | $400 |
| 441 | $441 = 7^2 \times 3^2$ P | $442 = 13 \times 17 \times 2$ N | | $441 |
| 483 | $483 = 3 \times 7 \times 23$ N | $484 = 11^2 \times 2^2$ P | | $484 |
| 511 | $511 = 7 \times 73$ N | $512 = 2^9$ P | | $512 |

| 216 | 200 | 175 | 169 | 160 | 150 | 144 | 125 |
|-----|-----|-----|-----|-----|-----|-----|-----|
| 225 | | | | | | | 100 |
| 243 | | | | | | | 90 |
| 288 | | THE | | | | | 81 |
| 289 | | DIVISOR | | | | | 72 |
| 330 | | GAME | | | | | 50 |
| 360 | | | | | | | 32 |
| GO | 4 | 6 | 10 | 12 | 18 | 25 | 30 |

**Fig. (19).** The divisor game.

## RULES FOR THE DIVISOR GAME (FIG.19)

1. Each player rolls die to determine order of turns and order of choice of banker.
2. The player tosses die and lands on number space. He/she must give both the correct prime factorization of the number and the number of divisors. For example, $36 = 2^2 \times 3^2$ and has 9 divisors: 1, 2, 3, 4, 6, 9, 12, 18, 36.

3. The player whose turn is next looks at back of number card. If original player is correct, he or she must multiply the exponents in the prime factorization together and then multiply by the number of divisors. If he/she is correct, player gets this amount of money from the bank. In the previous example, player would get (since $36 = 2^2$ x $3^2$ with 9 divisors) 2 x 2 x 9 = 36 dollars.
4. A hint to obtain the number of divisors without having to divide is to look at the relationship of the exponents in the prime factorization to the number of divisors.
5. When player lands on GO exactly, he/she receives $50 and game ends. If player passes GO, he/she proceeds to number space, takes turn, and then game ends. Player with most money wins the game.

## Number Cards for the Divisor Game

| 4 | $4 = 2^2$ | 1,2,4 = (3) | 2x3 = $6 |
|---|---|---|---|
| 6 | $6 = 2x3$ | 1, 2, 3, 6 = (4) | 1x1x4 = $4 |
| 10 | $10 = 2x5$ | 1,2, 5, 10 = (4) | 1x1x4 = $4 |
| 12 | $12 = 2^2x3$ | 1,2, 3,4, 6, 12 = (6) | 2x1x6 = $12 |
| 18 | $18 = 3^2x2$ | 1,2, 3, 6, 9, 18 = (6) | 2x1x6 = $12 |
| 25 | $25 = 5^2$ | 1,5, 25 = (3) | 1x2x3 = $6 |
| 30 | $30 = 3x2x5$ | 1, 2, 3, 5, 6, 10, 15, 30 = (8) | 1x1x1x8 = $8 |
| 32 | $32 = 2^5$ | 1,2,4, 8, 16, 32 = (6) | 5x6 = $30 |
| 50 | $50 = 5^2x2$ | 1,2, 5, 10, 25, 50 = (6) | 2x1x6 = $12 |
| 72 | $72 = 2^3x3^2$ | 1, 2, 3, 4, 6, 8, 9, 12, 18, 24, 36, 72 = (12) | 3x2x12 = $72 |
| 81 | $81 = 3^4$ | 1,3,9, 27, 81 = (5) | 4x5 = $20 |
| 90 | $90 = 3^2x2x5$ | 1, 2, 3, 5, 6, 9, 10, 15, 18, 30, 45, 90 = (12) | 2x1x1x12=$24 |
| 100 | $100 = 5^2x2^2$ | 1, 2, 4, 5, 10, 20, 25, 50, 100 = (9) | 2x2x9 = $36 |
| 125 | $125 = 5^3$ | 1,5, 25, 125 = (4) | 3x4 = $12 |
| 160 | $160 = 2^5x5$ | 1, 2,4, 8, 16, 32, 5, 10, 20,40, 80, 160 = (12) | 5x1x12 = $60 |
| 169 | $169 = 13^2$ | 1,13, 169 = (3) | 90, 120, 180, 360 = (24) |
| 175 | $175 = 5^2x7$ | 1,5, 7, 25, 35, 175 = (6) | 2x1x6 = $12 |
| 200 | $200 = 2^3x5^2$ | 1, 2, 4, 5, 8, 10, 20, 25, 40, 50, 100, 200 = (12) | 3x2x12 = $72 |
| 216 | $216 = 3^3x2^3$ | 1, 2, 3, 4, 6, 8, 9, 12, 18, 24, 27, 36, 56, 72, 108, 216 = (16) | 3x3x16 = $144 |
| 225 | $225 = 5^2x3^2$ | 1,3,5, 9, 15, 25, 45, 75, 225 = (9) | 2x2x9 = $36 |
| 243 | $243 = 3^5$ | 1, 3, 9, 27, 81, 243 = (6) | 5x6 = $30 |
| 288 | $288 = 2^5x3^2$ | 1, 2, 3, 4, 6, 8, 9, 12, 16, 18, 24, 32, 36, 48, 72, 96, 144, 288 = (18) | 5x2x18 = $180 |
| 289 | $289 = 17^2$ | 1,17, 289 = (3) | 2x3 = $6 |

| 330 | 330=11x3x2x5 | 1,2,3,5,6,10,11,15,20,30,33,55,66,110,165, 330 = (16) | 1x1x1x1x16 = $16 |
|---|---|---|---|
| 360 | 360 = $2^3$x$3^2$x5 | 1, 2, 3, 4, 5, 6, 8, 9, 10, 12, 15, 18, 20, 24, 30, 36, 40, 45, 60, 72, 90, 120, 180, 360 = (24) | 90, 120, 180, 360 = (24) |

| 73 | 71 | 67 | 61 | 59 | 53 | 47 | 43 |
|---|---|---|---|---|---|---|---|
| 79 | | | | | | | 41 |
| 83 | | | | | | | 37 |
| 89 | | THE | | | | | 31 |
| 97 | | SUM OF SQUARES GAME | | | | | 29 |
| 101 | | | | | | | 23 |
| 103 | | | | | | | 19 |
| GO | 2 | 3 | 5 | 7 | 11 | 13 | 17 |

**Fig. (20).** The sum of squares games (g).

## RULES FOR THE SUM OF SQUARES GAME (FIG.20)

1. Each player rolls die to determine order of turns and order of choice of banker.
2. The player rolls die and lands on number space. He/she must determine whether or not the prime number he/she landed on can be written as the sum of two squares. For example, 13 = 9 + 4, 9 = 3 x 3, 4 = 2 x 2, and thus 13 = $3^2$ + $2^2$, but 11 cannot be written as the sum of two squares.
3. The player whose turn is next looks at back of number card. If the number can be written as the sum of two squares and the original player answers correctly, he/she gets this amount of money from the bank. If the number cannot be written as the sum of two squares and the original player answers correctly, he/she must determine half of one less than the number space. If he/she determines this number correctly, player gets this amount of money from the bank.
4. Hint: There is a way to determine whether or not a number can be written as the sum of two squares, which involves prime factorization and the remainder when a number is divided by 4 (see Chapter 1).

5. If a player lands on GO exactly, he/she gets $50 and the game ends. If player passes GO, he/she does the number space. Player with the most money wins the game.

### Number Cards for the Sum of Squares Game

| $2$ | $2 = 1^2 + 1^2$ | $2 |
|---|---|---|
| 3 | No | $1 |
| 5 | $5 = 1^2 + 2^2$ | $5 |
| 7 | No | $3 |
| 11 | No | $5 |
| 13 | $13 = 2^2 + 3^2$ | $13 |
| 17 | $17 = 4^2 + 1^2$ | $17 |
| 19 | No | $9 |
| 23 | No | $11 |
| 29 | $29 = 5^2 + 2^2$ | $29 |
| 31 | No | $15 |
| 37 | $37 = 6^2 + 1^2$ | $37 |
| 41 | $41 = 5^2 + 4^2$ | $41 |
| 43 | No | $21 |
| 47 | No | $23 |
| 53 | $53 = 2^2 + 7^2$ | $53 |
| 59 | No | $29 |
| 61 | $61 = 5^2 + 6^2$ | $61 |
| 67 | No | $33 |
| 71 | No | $35 |
| 73 | $73 = 3^2 + 8^2$ | $73 |
| 79 | No | $39 |
| 83 | No | $41 |
| 89 | $89 = 5^2 + 8^2$ | $89 |
| 97 | $97 = 4^2 + 9^2$ | $97 |
| 101 | $101 = 1^2 + 10^2$ | $101 |
| $10^2$ | No | $51 |

| 21 | 20 | 19 | 18 | 17 | 16 | 15 | 14 |
|----|----|----|----|----|----|----|----|
| 22 | | | | | | | 13 |
| 23 | | | | | | | 12 |
| 24 | | | THE | | | | 11 |
| 25 | | | SYRACUSE ALGORITHM GAME | | | | 10 |
| 26 | | | | | | | 9 |
| 27 | | | | | | | 8 |
| GO | 1 | 2 | 3 | 4 | 5 | 6 | 7 |

**Fig. (21).** The syracuse algorithm game (e).

## RULES FOR THE SYRACUSE ALGORITHM GAME (FIG.21)

1. Each player rolls die to determine order of turns and order of choice of banker.
2. Player rolls die and lands on number space. He/she must write down all the numbers in the Syracuse Algorithm beginning with that number. The Syracuse Algorithm says: if a number is odd, multiply it by 3 and add 1; if a number is even, take half of it; continue the process until you reach 1.
3. The player whose turn is next looks at the back of the number card to see if the original player is correct. If the original player has all the correct numbers in the sequence, he/she gets the same number of dollars as how many numbers there are in the sequence—from the bank. (Note that for #27 only a sample of the numbers in the sequence are listed on the back of the card, since there are over 100 numbers in the sequence).
4. If a player lands on GO exactly, he/she receives $25 and the game ends. If a player passes GO, he/she goes through the procedure for the number space he/she landed on; then the game ends. The player with the most money is the winner.

### Number Cards for The Syracuse Algorithm Game

| 1 | 1-4-2-1 | $4 |
|---|---------|-----|
| 2 | 2-1 | $2 |
| 3 | 3-10-5-16-8-4-2-1 | $8 |
| 4 | 4-2-1 | $3 |
| 5 | 5-16-8-4-2-1 | $6 |

*Contd.....*

| 6 | 6-3-10-5-16-8-4-2-1 | $9 |
| 7 | 7-22-11-34-17-52-26-13-40-20-10-5-16-8-4-2-1 | $17 |
| 8 | 8-4-2-1 | $4 |
| 9 | 9-28-14-7-22-11-34-17-52-26-13-40-20-10-5-16-8-4-2-1 | $20 |
| 10 | 10-5-16-8-4-2-1 | $7 |
| 11 | 11-34-17-52-26-13-40-20-10-5-16-8-4-2-1 | $15 |
| 12 | 12-6-3-10-5-16-8-4-2-1 | $10 |
| 13 | 13-40-20-10-5-16-8-4-2-1 | $10 |
| 14 | 14-7-22-11-34-17-52-26-13-40-20-10-5-16-8-4-2-1 | $18 |
| 15 | 15-46-23-70-35-106-53-160-80-40-20-10-5-16-8-4-2-1 | $18 |
| 16 | 16-8-4-2-1 | $5 |
| 17 | 17-52-26-13-40-20-10-5-16-8-4-2-1 | $13 |
| 18 | 18-9-28-14-7-22-11-34-17-52-26-13-40-20-10-5-16-8-4-2-1 | $21 |
| 19 | 19-58-29-88-44-22-11 -34-17-52-26-13-40-20-10-5-16-8-4-2-1 | $21 |
| 20 | 20-10-5-16-8-4-2-1 | $8 |
| 21 | 21-64-32-16-8-4-2-1 | $8 |
| 22 | 22-11-34-17-52-26-13-40-20-10-5-16-8-4-2-1 | $16 |
| 23 | 23-70-35-106-53-160-80-40-20-10-5-16-8-4-2-1 | $16 |
| 24 | 24-12-6-3-10-5-16-8-4-2-1 | $11 |
| 25 | 25-76-38-19-58-29-88-44-22-11-34-17-52-26-13-40-20-10-5-16-8-4-2-1 | $23 |
| 26 | 26-13-40-20-10-5-16-8-4-2-1 | $11 |
| 27 | 27-82-41-124-62-31-94-47-142-71-214-107-321-161-484-242-121-.......1 | $112 |

**Fig. (22).** The fib-tri-prime game (f).

## RULES FOR THE FIB-TRI-PRIME GAME (FIG.22)

1. Roll die to determine the order of turns and order of choice of banker.
2. Player rolls die and lands on number space. He/she must say the corresponding Fibonacci number, Triangular number and prime number (see the Definitions below). In other words, if a player lands on 5, he/she must say the fifth Fibonacci number, fifth Triangular number, fifth prime number.
3. The player whose turn is next looks at the back of the appropriate number card to see if the original player is correct. If the player is correct, he/she multiplies the total number of digits in all three numbers by the number space he/she landed on. If he/she is again correct, he/she gets that amount of money from the bank.
4. If a player lands on GO exactly, he/she receives $100 and the game ends. If a player passes GO, he/she goes through the procedure for the number space he/she landed on; then the game ends. The player with the most money is the winner.

## DEFINITIONS FOR THE FIB-TRI-PRIME GAME

*Fibonacci numbers:* the numbers in the sequence 1, 1, 2, 3, 5, 8, 13, 21, 34, ...; the next number is obtained by adding the sum of the preceding number to the last number.

*Triangular numbers:* the numbers in the sequence 1, 3, 6, 10, 15, 21, 28, 36, ...; the next number is obtained by adding 1 more than the difference of the 2 preceding numbers to the last number.

*Prime Numbers:* 2, 3, 5, 7, 11, 13, 17, 19, ...; prime numbers are numbers whose only divisors are the number itself and 1. There is no sequence or pattern to prime numbers.

### Number Cards for the Fib-Tri-Prime Game

| 1 | F1 | T1 | P2 | $3 |
|---|-----|-----|-----|------|
| 2 | F1 | T3 | P3 | $6 |
| 3 | F2 | T6 | P5 | $9 |
| 4 | F3 | T10 | P7 | $16 |
| 5 | F5 | T15 | P11 | $25 |
| 6 | F8 | T21 | P13 | $30 |
| 7 | F13 | T28 | P17 | $42 |
| 8 | F21 | T36 | P19 | $48 |
| 9 | F34 | T45 | P23 | $54 |

*Contd.....*

| 10 | F55 | T55 | P29 | $60 |
|----|-----|-----|-----|-----|
| 11 | F89 | T66 | P31 | $66 |
| 12 | F144 | T78 | P37 | $84 |
| 13 | F233 | T91 | P41 | $91 |
| 14 | F377 | T105 | P43 | $112 |
| 15 | F610 | T120 | P47 | $120 |
| 16 | F987 | T136 | P53 | $128 |
| 17 | F1,597 | T153 | P59 | $153 |
| 18 | F2,584 | T171 | P61 | $162 |
| 19 | F4,181 | T190 | P67 | $171 |
| 20 | F6,765 | T210 | P71 | $180 |
| 21 | F10,946 | T231 | P73 | $210 |
| 22 | F17,711 | T253 | P79 | $220 |
| 23 | F28,657 | T276 | P83 | $230 |
| 24 | F46,368 | T300 | P89 | $240 |
| 25 | F75,025 | T325 | P97 | $250 |
| 26 | F121,393 | T351 | P101 | $312 |
| 27 | F196,418 | T378 | P103 | $324 |

# Games of Recreational Number Theory: Skill Levels Through Division and Fractions

**Abstract:** Chapter 4 consists of the games that require all the arithmetic skill levels through fractions. These games consist of The Clock Arithmetic Game, The Pascal's Triangle Game, The Anomalous Fractions Game, The Farey Fractions Game, and The Numberama Game. The games in Chapter 4 are appropriate in general for children in grades 6, 7, and 8, depending on their skill levels, inclusive of gifted children. These are the most challenging games in the set of Numberama Recreational Number Theory games in this book. These games can be especially stimulating for gifted children, and are an excellent way for middle school children to practice and improve upon their skills with fractions (see some of the teacher and teacher education Numberama workshop participant comments in Appendix). The Numberama Game is an excellent teaching device to practice virtually all of the Numberama Recreational Number Theory problems in Chapter 1 of this book.

**Keywords:** The anomalous fractions game, The clock arithmetic game, The farey fractions game, The numberama game, The Pascal's triangle game.

**Fig. (23).** The clock arithmetic game (h).

**Elliot Benjamin**

## RULES FOR THE CLOCK ARITHMETIC GAME (FIG.23)

1. Each player rolls die to determine order of turns and choice of banker.
2. He/she uses clock with the big number of hours and must try to find a number on the clock that when multiplied by the small number gives the number "1" on the clock; this number is called the "multiplicative inverse" of the small number. For example, number space $11_9$ means an 11-hour clock and $9 \times 5 = 45$ but going around the clock 4 times yields 1 since $11 \times 4 = 44$ and $45 = 44 + 1$, so 5 is the multiplicative inverse of 9 on an 11-hour clock: we will write $45 \equiv 1$ mod 4.
3. It may happen that there is no number which, when multiplied by the small number, yields 1. Look at $6_2$. For example, $2 \times 1 = 2$, $2 \times 2 = 4$, $2 \times 3 = 6$, $2 \times 4 = 8 \equiv 2$ mod 6; $2 \times 5 = 10 \equiv 4$ mod 6, $2 \times 6 = 12 \equiv 6$ mod 6, and the process just repeats. Two, therefore, has no mulitplicative inverse on a 6-hour clock.
4. Player must say what the multiplicative inverse is of the small number on the big number clock—or state there is no multiplicative inverse if this is the case.

For another example, take $8_5 = 5 \times 5 = 25$, $25 = 24 + 1$, and $24 = 8 \times 3$ goes around an 8-hour clock 3 times so $25 = 1$ mod 8; 8 is therefore the multiplicative inverse of 5 on an 8-hour clock.

5. Player whose turn is next looks at back of number card. If original player is correct that there is no multiplicative inverse, he/she gets the amount of money from the bank of the big number on the number space. If the player gives the correct multiplicative inverse, he/she must multiply the multiplicative inverse by the clock number (big number). For example, on number space $11_9$, since the multiplicative inverse of 9 is 5, player would get $11 \times 5 = \$55$ from the bank.
6. If player lands on GO exactly, he/she gets \$100 and game ends. If player passes GO, he/she does procedure on number space, then game ends; player with the most money wins the game.

### Number Cards for the Clock Arithmetic Game

| | | |
|---|---|---|
| $3_2$ | $2 \times 2 = 4 \equiv 1$ mod 3 | $2 \times 3 = \$6$ |
| $4_3$ | $3 \times 3 = 9 \equiv 1$ mod 4 | $3 \times 4 = \$12$ |
| $5_3$ | $3 \times 2 = 6 \equiv 1$ mod 5 | $2 \times 5 = \$10$ |
| $6_3$ | No | \$6 |
| $7_6$ | $6 \times 6 = 36 \equiv 1$ mod 7 | $6 \times 7 = \$42$ |
| $8_5$ | $5 \times 5 = 25 \equiv 1$ mod 8 | $5 \times 8 = \$40$ |
| $9_6$ | No | \$9 |
| $10_4$ | No | \$10 |
| $11_9$ | $9 \times 5 = 45 \equiv 1$ mod 11 | $5 \times 11 = \$55$ |

*Contd.....*

| | | |
|---|---|---|
| $12_5$ | $5 \times 5 = 25 \equiv 1 \bmod 12$ | $5 \times 12 = \$60$ |
| $13_7$ | $7 \times 2 = 14 \equiv 1 \bmod 13$ | $2 \times 13 = \$26$ |
| $14_3$ | $3 \times 5 = 15 \equiv 1 \bmod 14$ | $5 \times 14 = \$70$ |
| $15_5$ | No | $\$15$ |
| $16_3$ | $3 \times 11 = 33 \equiv 1 \bmod 16$ | $11 \times 16 = \$176$ |
| $17_2$ | $2 \times 9 = 18 \equiv 1 \bmod 17$ | $9 \times 17 = \$153$ |
| $18_{11}$ | $11 \times 5 = 55 \equiv 1 \bmod 18$ | $18 \times 5 = \$90$ |
| $19_5$ | $5 \times 4 = 20 \equiv 1 \bmod 19$ | $4 \times 19 = \$76$ |
| $20_7$ | $7 \times 3 = 21 \equiv 1 \bmod 20$ | $3 \times 20 = \$60$ |
| $21_3$ | No | $\$21$ |
| $22_9$ | $9 \times 5 = 45 \equiv 1 \bmod 22$ | $5 \times 22 = \$110$ |
| $23_{18}$ | $18 \times 9 = 162 \equiv 1 \bmod 23$ | $9 \times 23 = \$207$ |
| $24_7$ | $7 \times 7 = 49 \equiv 1 \bmod 24$ | $7 \times 24 = \$168$ |
| $25_6$ | $6 \times 21 = 126 \equiv 1 \bmod 25$ | $21 \times 25 = \$525$ |
| $26_3$ | $3 \times 9 = 27 \equiv 1 \bmod 26$ | $9 \times 26 = \$234$ |
| $27_{10}$ | $10 \times 19 = 190 \equiv 1 \bmod 27$ | $19 \times 27 = \$513$ |
| $28_{11}$ | $11 \times 23 = 253 \equiv 1 \bmod 28$ | $23 \times 28 = \$644$ |
| $29_{28}$ | $28 \times 28 = 784 \equiv 1 \bmod 29$ | $28 \times 29 = \$812$ |

**Fig. (24).** The pascal's triangle game (h).

## RULES FOR THE PASCAL'S TRIANGLE GAME (FIG.24)

1. Each player rolls die to determine order of turns and order of choice of banker.
2. Player rolls die and lands on number space. He/she must give numbers of the corresponding row of Pascal's Triangle. Pascal's Triangle is the following:

1

1 1

1 2 1

1 3 3 1

1 4 6 4 1

1 5 10 10 5 1

1 6 15 20 15 6 1

The top row is called row 0 and each row is obtained by putting ones on the outside and adding together the two adjacent numbers in the row, writing the sum halfway below the two numbers in the next row. If row 5 is 1 5 10 10 5 1, row 6 will be 1 6 15 20 15 6 1.

3. If the number space says letter A next to the number, player must add together all the entries in the row. Thus, for row 6, the sum is $1 + 6 + 15 + 20 + 15 + 6 + 1 = 64$. If the number space says letter B, player must divide each entry in the row, other than 1, by the row number, and decide if it goes into the row entry evenly. Thus, for row 5, 5 goes into all the row entries, other than 1, evenly—as 5 divides evenly into 5 and 10. However, for row 6, 6 divides evenly only into 6, as 6 does not go evenly into 15 and 20.

4. The player whose turn is next looks at back of number card. If original player is correct, he/she gets the largest number in the row as money from the bank. Thus for row 5, player would get $10; for row 6, player would get $20.

5. If player lands on GO exactly, he/she gets $200 and game ends. If player passes GO, he/she does procedure for number space; then game ends. Player with the most money is the winner.

### Number Cards for the Pascal's Triangle Game

| Row 0 | 1 |
|-------|---|

*Contd.....*

| Row 1 | 1 1 |
|---|---|
| Row 2 | 1 2 1 |
| Row 3 | 1 3 3 1 |
| Row 4 | 1 4 6 4 1 |
| Row 5 | 1 5 10 10 5 1 |
| Row 6 | 1 6 15 20 15 6 1 |
| Row 7 | 1 7 21 35 35 21 7 1 |
| Row 8 | 1 8 28 56 70 56 28 8 1 |
| Row 9 | 1 9 36 84 126 126 84 36 9 1 |
| Row 10 | 1 10 45 120 210 252 210 120 45 10 1 |
| Row 11 | 1 11 55 165 330 462 462 330 165 55 11 1 |
| Row 12 | 1 12 66 220 495 792 924 792 495 220 66 12 1 |
| Row 13 | 1 13 78 286 715 1287 1716 1716 1287 715 286 78 13 1 |
| Row 14 | 1 14 91 364 1001 2002 3003 3432 3003 2002 1008 364 91 14 1 |
| Row 15 | 1 15 105 455 1365 3003 5005 6435 6435 5005 3003 1365 455 105 15 1 |

| | | |
|---|---|---|
| 2A | $1+2+1 = 4$ | $2 |
| 3A | $1+3+3+1 = 8$ | $8 |
| 3B | 3 - Yes | $8 |
| 4A | $1+4+6+4+1 = 16$ | $6 |
| 4B | 4 - Yes; 6 - No | $6 |
| 5A | $1+5+10+10+5+1 = 32$ | $10 |
| 5B | 5, 10 - Yes | $10 |
| 6A | $1+6+15+20+15+6+1 = 64$ | $20 |
| 6B | 6 - Yes; 15, 20 - No | $20 |
| 7A | $1+7+21+35+35+21+7+1 = 128$ | $35 |
| 7B | 7, 21, 35 - Yes | $35 |
| 8A | $1+8+28+56+70+56+28+8+1 = 256$ | $70 |
| 8B | 8, 56 - Yes; 28, 70 - No | $70 |
| 9A | $1+9+36+84+126+126+84+36+9+1 = 512$ | $126 |
| 9B | 9, 36, 126 - Yes; 84 - No | $126 |
| 10A | $1+10+45+120+210+252+210+120+45+10+1 = 1024$ | $252 |
| 10B | 10, 120, 210 - Yes; 45,252 - No | $252 |
| 11A | $1+11+55+165+330+462+462+330+165+55+11+1 = 2048$ | $462 |
| 11B | 11, 55, 165, 330, 462 - Yes | $462 |
| 12A | $1+12+66+220+495+792+924+792+495+220+66+12+1 = 4096$ | $924 |
| 12B | 12, 792,924 - Yes; 66,220,495 - No | $924 |
| 13A | $1+13+78+286+715+1287+1716+1716+1287+715+286+78+13+1 = 8192$ | $1716 |
| 13B | 13, 78,286, 715, 1237, 1716 - Yes | $1716 |
| 14A | $1+14+91+364+1001+2002+3003+3432+3003+2002+1001+364+91+14+1 = 16384$ | $3003 |
| 14B | 14, 364, 2002 - Yes; 91, 1001, 3003, 3432 - No | $3003 |

*Contd.....*

| 15A | 1+15+105+455+1365+3003+5005+6435+6435+5005+3003+1365+3003+5005+6435+6435+5005+3003+1365+455+105+15+1 = 32,768 | $6435 |
|-----|---|---|
| 15B | 15, 105, 1365, 6435 - Yes; 455, 3003, 5005 - No | $6435 |

$$\frac{37}{74} \quad \frac{28}{86} \quad \frac{36}{68} \quad \frac{22}{28} \quad \frac{49}{98} \quad \frac{48}{88} \quad \frac{36}{66} \quad \frac{24}{48}$$

$$\frac{15}{54} \qquad\qquad\qquad\qquad\qquad \frac{39}{94}$$

$$\frac{24}{46} \qquad\qquad\qquad\qquad\qquad \frac{19}{95}$$

$$\frac{26}{64} \qquad \textbf{THE} \qquad \frac{28}{88}$$

$$\textbf{ANOMALOUS}$$
$$\textbf{FRACTIONS}$$
$$\frac{26}{65} \qquad \textbf{GAME} \qquad \frac{17}{71}$$

$$\frac{44}{48} \qquad\qquad\qquad\qquad\qquad \frac{46}{68}$$

$$\frac{38}{86} \qquad\qquad\qquad\qquad\qquad \frac{14}{48}$$

$$\textbf{GO} \quad \frac{18}{84} \quad \frac{26}{66} \quad \frac{35}{55} \quad \frac{14}{42} \quad \frac{16}{64} \quad \frac{38}{84} \quad \frac{23}{39}$$

Fig. (25). The anomalous fractions game (h).

## RULES FOR THE ANOMALOUS FRACTIONS GAME (FIG.25)

1. Each player rolls die to determine order of turns and order of choice of banker.
2. Player rolls die and lands on number space. He/she reduces fractions to lowest terms, then cancels the right-hand digit in the numerator and left-hand digit in the denominator of original fraction, and reduces this new fraction to lowest terms. For example, if player lands on fraction 24/48, 24/48 = 12/24 = 6/12 = 1/2 and 24/48 = 2/8 = 1/4.

If these two resulting fractions are the same, the original fraction is said to be "anomalous." Thus 24/48 is not anomalous.

3. The player then multiplies numerator of one fraction by denominator of other fraction, and subtracts smaller number from larger. For 24/48, we get the two fractions 1/4 and 1/2, and 1 x 4 − 1 x 2 = 2. If the original fraction is anomalous, then, since the two fractions are the same, we would end up with zero. Player then subtracts this final number from 50. Thus for 24/48, we would have 50 − 2 = 48. If final number is larger than 50, player does not get any money from bank.
4. The player whose turn is next looks at back of number card. If both fractions are correctly reduced to lowest terms and final number is correct, player receives

final number subtracted from 50 amount of money from bank, if original fraction is not anomalous. If original fraction is anomalous and player has the correctly reduced fractions, he/she receives $200, if she/he states that the fractions are anomalous; if this is not stated but the fractions and final number are correct, player receives $50.

5.  If player lands on GO exactly, he/she receives $50 and game ends. If player passes GO, he/she does number space procedure and then game ends. The player with the most money wins the game.

### Number Cards for the Anomalous Fractions Game

| | | | | |
|---|---|---|---|---|
| 18/84 | 18/84 = 1/4 | 18/84 = 9/42 = 3/14 | 1x14 - 4x3 = 2 | $48 |
| 26/66 | 26/66 = 2/6 = 1/3 | 26/66 = 13/33 | 13/3 - 33x1 = 6 | $44 |
| 35/55 | 35/55 = 3/5 | 35/55 = 7/11 | 7x5 - 11x3 = 2 | $48 |
| 14/42 | 14/42 = 1/2 | 14/42 = 7/21 = 1/3 | 1x3 - 1x2 = 1 | $49 |
| 16/64 | 16/64 = 1/4 | 16/64 = 8/32 = 1/4 | 1x4 - 4x1 = 0 | Anomalous $200 |
| 38/84 | 38/84 = 3/4 | 38/84 = 19/42 | 42x3 - 19x4 = 50 | $0 |
| 14/48 | 14/48 = 1/8 | 14/48 = 7/24 | 8x7 - 1x24 = 32 | $18 |
| 46/68 | 46/68 = 4/8 = 1/2 | 46/68 = 23/34 | 23x2 - 1x34 = 12 | $38 |
| 23/39 | 23/39 = 2/9 | 23/39 = 23/39 | 23x9 - 2x39 = 129 | $0 |
| 17/71 | 17/71=1/1 = 1 | 17/71 = 17/71 | 1x71 - 1x17 = 54 | $0 |
| 28/88 | 28/88 = 2/8 = 1/4 | 28/88 = 14/44 = 7/22 | 7x4 - 1x22 = 6 | $44 |
| 19/95 | 19/95 = 1/5 | 19/95 = 1/5 | 1x5 - 5x1 = 0 | Anomalous $200 |
| 39/94 | 39/94 = 3/4 | 39/94 = 39/94 | 94x3 - 39x4 = 126 | $0 |
| 24/48 | 24/48 = 2/8 = 1/4 | 24/48 = 12/24 = 6/12 = 1/2 | 1x4 – 1x2 = 2 | $48 |
| 36/66 | 36/66 = 3/6 = 1/2 | 36/66 = 18/33 = 6/11 | 6x2 - 1x11 = 1 | $49 |
| 48/88 | 48/88 = 4/8 = 1/2 | 48/88 = 24/44 = 12/22 = 6/11 | 6x2 – 1x11 = 1 | $49 |
| 49/98 | 49/98 = 4/8 = 1/2 | 49/98 = 7/14 = 1/2 | 1x2 - 2x1 = 0 | Anomalous $200 |
| 22/28 | 22/28 = 2/8 = 1/4 | 22/28 = 11/14 | 11x4 – 14 x1 = 30 | $20 |
| 36/68 | 36/68 = 3/8 | 36/68 = 18/34 = 9/17 | 9x8 - 17x3 = 21 | $29 |
| 28/86 | 28/86 = 2/6 = 1/3 | 28/86 = 14/43 | 43x1 - 14x3 = 1 | $49 |
| 37/74 | 37/74 = 3/4 | 37/74 = 1/2 | 3x2 - 1x4 = 2 | $48 |
| 15/54 | 15/54 = 1/4 | 15/54 = 5/18 | 5x4 - 1x18 = 2 | $48 |
| 24/46 | 24/46 = 2/6 = 1/3 | 24/46 = 12/23 | 12x3 - 23x1 = 13 | $37 |
| 26/64 | 26/64 = 2/4 = 1/2 | 26/64 = 13/32 | 32x1 - 13x2 = 6 | $44 |
| 26/65 | 26/65 = 2/5 | 26/65 = 2/5 | 5x2 - 5x2 = 0 | Anomalous $200 |
| 44/48 | 44/48 = 4/8 = 1/2 | 44/48 = 11/12 | 11x2 - 12x1 = 10 | $40 |
| 38/86 | 38/86 = 3/6 = 1/2 | 38/86 = 19/43 | 43x1 - 19x2 = 5 | $45 |

Board (reading around the outer ring):

Top row: 8/1, 7/17, 7/14, 7/11, 7/8, 7/7, 7/3, 6/13

Left column (top to bottom): 8/1, 8/4, 8/9, 8/12, 8/15, 8/17, 8/20, GO

Right column (top to bottom): 6/13, 6/12, 6/9, 6/7, 6/4, 5/9, 5/5, 5/3

Bottom row: GO, 3/2, 3/3, 3/4, 4/3, 4/5, 4/6, 5/3

Center: **THE FAREY FRACTIONS GAME**

**Fig. (26).** The farey fraction's game (h).

## RULES FOR THE FAREY FRACTIONS GAME (FIG. 26)

1. Each player rolls die to determine order of turns and order of choice of banker.
2. The player rolls die and lands on number space. The big number refers to the sequence of Farey fractions of a given order; the small number refers to the fraction in the sequenced ordering from smallest to largest.

Farey fractions of order n are defined to be all proper fractions between 0 and 1, reduced to lowest terms, in increasing order, with a denominator less than or equal to n. For example, the Farey fractions of order 5 are:

0/1, 1/5, 1/4, 1/3, 2/5, 1/2, 3/5, 3/4, 4/5, 1/1; $5_7$ would be 3/5. Note: any two successive Farey fractions have the property that when they are cross-multiplied their difference will be 1. For example, 3/5 and 2/3: 5x2 − 3x3 = 1; 2/3 and 3/4: 3x3 − 2x4 = 1; 0/1 and 1/5: 1x1 − 0x5 = 1.

3. The player whose turn is next looks at back of number card to see if original player is correct. If player is correct, he/she must then correctly state the product of the Farey order with the fraction order, *i.e.*, the product of the big number and small number on the number space; for 57 this product would be 5 x 7 = 35. If correct, player gets this amount of money from bank.

4.  If player lands on GO exactly, he/she gets $50 and then game ends. If player passes GO, he/she lands on number space, does procedure, and then game ends. The player with the most money is the winner.

### Number Cards for the Farey Fractions Game

| $3_2$ | 1/3 | $6 |
|---|---|---|
| $3_3$ | 1/2 | $9 |
| $3_4$ | 2/3 | $12 |
| $4_3$ | 1/3 | $12 |
| $4_5$ | 2/3 | $20 |
| $4_6$ | 3/4 | $24 |
| $5_3$ | 1/4 | $15 |
| $6_5$ | 2/5 | $25 |
| $5_9$ | 3/4 | $45 |
| $6_4$ | 1/4 | $24 |
| $6_7$ | 1/2 | $42 |
| $6_9$ | 2/3 | $54 |
| $6_{11}$ | 4/5 | $66 |
| $6_{12}$ | 5/6 | $78 |
| $7_3$ | 1/6 | $21 |
| $7_6$ | 2/7 | $42 |
| $7_8$ | 2/5 | $56 |
| $7_{11}$ | 4/7 | $77 |
| $7_{14}$ | 3/4 | $98 |
| $7_{17}$ | 6/7 | $119 |
| $8_1$ | 0/1 | $8 |
| $8_4$ | 1/6 | $32 |
| $8_9$ | 3/8 | $72 |
| $8_{12}$ | 1/2 | $96 |
| $8_{15}$ | 5/8 | $120 |
| $8_{17}$ | 3/4 | $136 |
| $8_{20}$ | 7/8 | $160 |

| Syracuse Algorithm | Number of Divisors | Sums of Numbers | Number of Subsets | Powerful Numbers | Amicable Numbers | Weird Numbers | Perfect Numbers |
|---|---|---|---|---|---|---|---|
| Multiplicative Persistence | | | | | | | Semi-Perfect Numbers |
| Anomalous Fractions | | | | | | | Twin-Prime Numbers |
| Pascal's Triangle | | **THE** | | | | | Kaprekar Numbers |
| Clock Arithmetic | | **NUMBERAMA** | | | | | Semi-Prime Numbers |
| Farey Fractions | | **GAME** | | | | | Odd Abundant Numbers |
| Magic Squares | | | | | | | Even Abundant Numbers |
| **GO** | Counting Numbers | Even Counting Numbers | Odd Counting Numbers | Multiples of 7 | Triangular Numbers | Fibonacci Numbers | Prime Numbers |

**Fig. (27).** The numberama game (h).

## RULES FOR THE NUMBERAMA GAME (FIG. 27)

1. Numberama may be played by a maximum of 6 players. Each player rolls the die to determine order of turns and order of choice of banker.

2. When a player lands on a number space, he/she rolls die again. The player who has the following turn takes the number card and asks the player who just rolled the die the question on the card, telling him/her the number of dollars for each question.

3. A player has up to 5 minutes to come up with an answer to a particular question on the number card. Calculators may be used to come up with answers only if all players have agreed to this at the start of the game.

4. The player who rolled the die has the option of looking at the Definitions, Examples, Hints section in the Appendix to help come up with the correct answer.

5. Most often the correct answers are completely listed on the number card and the player who is reading the card will state which answer is correct or not. If the answer is not listed on the number card, then all the other players in the game must agree that the player who rolled the die has come up with the correct answer.

6. To obtain the designated amount of money from the bank, the player who rolled the die must give a complete answer to a particular question on the number card.

7. If a player lands on GO exactly, he/she gets $200 and the game ends. If a player

passes GO, he/she does the procedure on the number space and the game ends. The player with the most money is the winner.

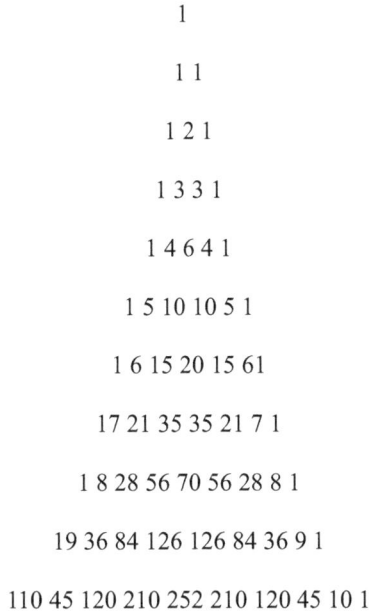

1

1 1

1 2 1

1 3 3 1

1 4 6 4 1

1 5 10 10 5 1

1 6 15 20 15 61

17 21 35 35 21 7 1

1 8 28 56 70 56 28 8 1

19 36 84 126 126 84 36 9 1

110 45 120 210 252 210 120 45 10 1

**Fig. (28).** Pascal's triangle in number cards for the numberama game.

## NUMBER CARDS FOR THE NUMBERAMA GAME (FIG.28)

1. Counting Numbers: roll die: pay number of dollars if player can say the first twelve counting numbers: 1, 2, 3, 4, 5, 6, 7, 8, 9, 10, 11, 12.
2. Even Counting Numbers: roll die: pay twice number of dollars if player can say first twelve even counting numbers (starting with 2): 2, 4, 6, 8,10, 12, 14, 16, 18, 20, 22, 24.
3. Odd Counting Numbers: roll die: pay three times number of dollars if player can say the first twelve odd counting numbers: 1, 3, 5, 7, 9, 11, 13, 15, 17, 19, 21, 23.
4. Multiples of 7: roll die: pay four times the number of dollars if player can say the first twelve multiples of 7: 7, 11, 21, 28, 28, 35, 42, 49, 56, 63, 70, 77, 84.
5. Triangular Numbers: roll die: pay five times the number of dollars if player can say the first twelve triangular numbers: 1, 3, 6, 10, 15, 21, 28, 36, 45, 55, 66, 78.
6. Fibonacci Numbers: roll die: pay six times the number of dollars if player can say the first twelve Fibonacci numbers: 1, 1, 2, 3, 5, 8, 13, 21, 34, 55, 89, 144.
7. Prime Numbers: roll die, pay ten times the number of dollars if player can say the first twenty prime numbers: 2, 3, 5, 7, 11, 13, 17, 19, 23, 29, 31, 37, 41,

43, 47, 53, 59, 61, 71, 73.

8. Even Abundant Numers: roll die: pay twelve times the number of dollars if player can say the first six even abundant numbers, and show that they are abundant: 12, 18, 20, 24, 30, 36.

9. Odd Abundant Numbers: roll die: pay 10 times the number of dollars if player can say an odd abundant number and show that it is an odd abundant number. The smallest odd number is 945; its proper divisors are: 1, 3, 315, 5, 189, 7, 135, 9, 105, 15, 63, 21, 45, 27, and 35; the proper divisors add up to 975, so 945 is abundant.

10. Semi-Prime Numbers: roll die: pay 10 times the number of dollars if player can say three consecutive semi-prime numbers and show that they are semi-prime. Pay 15 times the number of dollars if player can do this for a second set of three consecutive semi-prime numbers. The two sets are 33, 34, 35; 85, 86, 87.

11. Kaprekar Numbers: roll die: pay 10 times the number of dollars if player can say one Kaprekar number greater than 50 and show it is Kaprekar. Pay 15 times the number of dollars if player can do this for a second Kaprekar number greater than 50. The two Kaprekar numbers greater than 50 are 55; $55^2$ = 3025, and 30 + 25 = 55; 99; $99^2$ = 9801, and 98 + 1 = 99.

12. Twin-Prime Numbers: roll die: pay 14 times the number of dollars if player can say the first 20 twin-prime numbers: 2, 3, 5, 7, 11, 13, 17, 19, 29, 31, 41, 43, 59, 61, 71, 73, 101, 103, 107, 109.

13. Semi-Perfect Numbers: roll die: pay 15 times the number of dollars if player can show that 20, 40, 60, 80, and 100 are semi-perfect numbers. Some possible answers are: 20 = 1 + 4 + 5 + 10; 40 = 20 + 10 + 5 + 4 + 1; 60 − 30 + 20 + 10; 80 = 40 + 20 + 10 + 5 + 4 + 1; 100 = 50 + 25 + 20 + 5.

14. Perfect Numbers: roll die: pay 15 times the number of dollars if player can say the first five perfect numbers: 6; 28; 496; 8,128; 33,550,336; or pay 25 times the number of dollars if player can also show that the first three perfect numbers are perfect: 6 = 1 + 2 + 3; 28 = 1 + 2 + 4 + 7 + 14; 496 = 1 + 2 + 248 + 4 + 124 + 8 + 62 + 16 + 31; or pay 40 times the number of dollars if player can also show that the fourth perfect number is perfect: 8,128 = 1 + 2 + 4064 + 4 + 2032 + 8 + 1016 + 16 + 508 + 32 + 254 + 64 + 127; or pay 100 times the number of dollars if player can also show that the fifth perfect number is perfect 33,550,336 = ? All players need to check this answer carefully.

15. Weird Numbers: roll die: pay 25 times the number of dollars if player can say a two-digit "weird" number and show that it is "weird": the answer is 70, since the proper divisors of 70 are 1, 2, 35, 5, 14, 10, and 7, and no combination of these numbers add up to 70, but their sum is 74. Seventy is abundant but not semi-perfect, or "weird."

16. Amicable Numbers: roll die: pay 30 times the number of dollars if player can

say a pair of amicable numbers and show that they are amicable. The smallest pair of amicable numbers is 220 and 284, as the proper divisors of 220 are: 1, 2, 110, 4, 55, 10, 22, 11, 20, 4, 44, and these numbers add up to 284; the proper divisors of 284 are: 1, 2, 142, 4, 71, and these numbers add up to 220.

17. Powerful Numbers: roll die: pay 20 times the number of dollars if player can say two consecutive three-digit powerful numbers and show that they are powerful. The answer is 288 and 289, since $288 = 32 \times 9 = 2^5 \times 3^2$, and $289 = 17 \times 17 = 17^2$.

18. Number of Subsets: roll die: pay 20 times the number of dollars if player can list all subsets in a set of five numbers: (1, 2, 3, 4, 5) —there are 32 subsets—or pay 30 times the number of dollars if player can also list all subsets in a set of six numbers: (1, 2, 3, 4, 5, 6)—there are 64 subsets.

19. Sums of Squares (for prime numbers): roll die: pay 15 times the number of dollars if player can say three prime numbers greater than 100 that are the sum of two squares, and show how they are the sum of two squares. Pay 20 times the number dollars if player can do this for four prime numbers greater than 100; 25 times the number of dollars for five prime numbers greater than 100. The five smallest prime numbers greater than 100 that can be written as the sum of two squares are: $101 = 1^2 + 10^2$; $109 = 3^2 + 10^2$; $113 = 8^2 + 7^2$ ; $137 = 11^2 + 4^2$; $149 = 7^2 + 10^2$.

20. Number of Divisors: roll die: pay 20 times the number of dollars if player can say the number of divisors for 100, 200, 300, 400, and 500. It is not necessary to say what the actual divisors are. A hint is to use the exponents in their prime factorizations. The answers are $100 = 2^2 \times 52$ has 9 (3 x 3) divisors; $200 = 2^3 \times 5^2$ has 12 (4 x 3) divisors; $300 = 2^2 \times 52 \times 3^1$ has 18 (3 x 3 x 2) divisors; $400 - 2^4 \times 5^2$ has 15 (5 x 3) divisors; $500 - 2^2 \times 5^3$ has 12 (3 x 4) divisors.

21. Syracuse Algorithm: roll die: pay 15 times the number of dollars if player can say all numbers in Syracuse Algorithm for 19: 19, 58, 29, 88, 44, 22, 11, 34, 17, 52, 26, 13, 40, 20, 10, 5, 16, 8, 4, 2, 1; or pay 40 times the number of dollars if player can say all numbers in Syracuse Algorithm for 27: 27, 82, 41, 124, 62, 31, 94, 47, 142, 71, 214, 107, 332.... (It takes 112 numbers to reach 1).

22. Multiplicative Persistence: roll die: pay 15 times the number of dollars if player can say three numbers, each having multiplicative persistence of 3, and show that this is the case; or pay 25 times the number of dollars if player can say a number of multiplicative persistence 4 and show that this is the case (only two-digit numbers may be used). Three numbers that have multiplicative persistence of 3 are 87, 97, and 68. All players should correctly verify the only two-digit number with multiplicative persistence of 4.

23. Anomalous Fractions: roll die: pay 15 times the number of dollars if player can say two anomalous fractions; or pay 20 times the number of dollars if

player can say three anomalous fractions; or pay 25 times the number of dollars if player can say four anomalous fractions. The four anomalous fractions are: $16/64 = 1/4$; $19/95 = 1/5$; $49/98 - 4/8 = 1/2$; $26/65 = 2/5$.

24. Pascal's Triangle: roll die: pay 15 times the number of dollars if player can construct the first 12 rows of Pascal's Triangle (starting with row 0); or pay 30 times the number of dollars if player can say which row numbers divide every entry—other than 1—in the row. The answer to this last question is: rows 2, 3, 5, 7, 11; in general it works only for prime number rows:

25. Clock Arithmetic: roll die: pay 30 times the number of dollars if player can construct all clocks from a 2-hour clock to a 12-hour clock, and correctly show which clocks have multiplicative inverses for each of their numbers, and say what the multiplicative inverses are. Clocks numbers 2, 3, 5, 7 and 11 have multiplicative inverses for each of their numbers; in general this works only for prime number clocks.

26. Farey Fractions: roll die: pay 15 times the number of dollars if player can say all fractions in correct order in Farey sequence of Order 6; or pay 20 times the number of dollars if player can do this for Farey sequence of Order 7; or pay 30 times the number of dollars if player can do this for Farey sequence of Order 8.
    Order 6: 0/1, 1/6, 1/5, 1/4, 1/3, 2/5,1/2, 3/5, 2/3, 3/4, 4/5, 5/6, 1/1
    Order 7: 0/1, 1/7, 1/6, 1/5, 1/4, 2/7, 1/3, 2/5, 3/7, 1/2, 4/7, 3/5, 2/3, 5/7, 3/4, 4/5, 5/6, 6/7, 1/1
    Order 8: 0/1, 1/8, 1/7, 1/6, 1/5, 1/4, 2/7, 1/3, 3/8, 2/5, 3/7, 1/2, 4/7, 3/5, 5/8, 2/3, 5/7, 3/4, 4/5, 5/6, 6/7, 7/8, 1/1

27. Magic Squares: roll die: pay 30 times the number of dollars if player can construct a 3 by 3 magic square, using all numbers from 1 to 9 exactly once, with a magic number of 15; or pay 50 times the number of dollars if player can construct a 4 by 4 magic square, using all numbers from 1 to 16 exactly once, with a magic number of 34; or pay 75 times the number of dollars if player can construct a 5 by 5 magic square, using all numbers from 1 to 25 exactly once, with a magic number of 65. All players must verify that these magic squares are indeed correct.

# Appendix 1: Participants' Responses to Numberama Program

## APPENDIX 1 ABSTRACT AND KEYWORDS

In this Appendix 1, I describe a variety of participants' responses to my Numberama program. The participants that I have worked with in my Numberama program include regular elementary school and middle school students, gifted elementary school and middle school students, elementary school teachers, elementary school parents, elementary school workshop participants, gifted children's workshop participants, children at a mental hospital, retirement home residents, various family members and friends, and college students in my introductory mathematics and psychology classes. Unfortunately, I do not have written records of the responses to all my Numberama activities. However, the following sample of participants' responses that I am including in this Appendix is I believe a testimony to the tremendous value of my Numberama program. From my lifelong experience of engaging Numberama with children, I conclude that the greatest benefit of Numberama is for gifted elementary school children in grades 4 – 6, though as can be seen from the student and teacher responses below, Numberama has also had much value for children in regular elementary school classes, especially in grades 5 and 6. I have also learned, as described below, that Numberama has much value in college teaching, both in teacher education classes and in liberal mathematics classes. The following sections in part of the Appendix describe participants' responses to Numberama in a number of diverse settings.

### Appendix 1 Keywords

1. Gifted children
2. Mathworks
3. Finite Math
4. Teacher education classes
5. Family Math
6. Retirement homes
7. Elementary school regular classes

## NUMBERAMA AT AN ELEMENTARY SCHOOL AS ENRICHMENT BY CHOICE: 2007

### Overview

In 2007 I was employed as a mathematics consultant to work in a school system in Maine with $4^{th}$, $5^{th}$, and $6^{th}$ grade students who chose to engage in math enrichment activities. I presented various Numberama Recreational Number Theory problems described in Chapter 1 in this book for 15 one hour sessions every other week, and in one session the children played some of the Numberama games in Chapters 2, 3, and 4. The responses of the children were exceptionally enthusiastic, as they welcomed Numberama as a refreshing change of pace from their usual mathematics lessons. Furthermore, a number of them expressed that they would have liked to have Numberama more often than every other week and for longer sessions, as they would forget material from one session to the next. The following are the responses I received to two questions: 1) What did you like most about the program? and 2) What would you change about the program?

## What Did You Like Most About the Program?

–I liked the whole thing.
–I liked learning new things because it helped me in my normal class. . . like the factor tree.
–I how we talk about things and he lets us try to find the patterns before he tells us. I think it is worth paying for.
I liked the people in it. I liked how I got to learn more ways to do my mathematics and I liked what we did in Numberama.
–I like the game that we played and next year have more games please.
–I thought it was pretty cool that you could do such a simple pattern to get an accurate answer.
–I <u>LOVE</u> everything. It's really fun being here with all these kids and doing math! I love math. My friends think that I am crazy that I do. I was K. . . 's math partner and we both went to Numberama. It helped us a lot in our other math and made it quicker and easier.
–I liked the part of this where we learned about prime numbers and when we played the circle game.
–I liked it because it gave me a chance to learn something new.
–I really like math and I really liked the challenge of the Numberama program. I think that it should be continued every year. It's really worth the money.
–I like how we talk about different things and learn different things.

## What Would You Change About the Program?

–I wouldn't change anything.
–More games.
–I think we should have a longer time or on more days.
–I would change that we have it every week just like we have our specials. I think the 5th and 6th graders should do it together then after the 3rd and 4th could to it together.
–I wouldn't change anything really about Numberama. I would only say that we should have a little more time.
–Keep it the same.
–I think we should do Numberama longer.
–I do wish the classes were closer together because it is hard to remember what we did last class.
–I think the classes should be longer and they should be more than every other week. Other than that I really like the class.
–I wish we could have a group of 5th and 6th grades so it's a little more challenging and one for 3rd and 4th grades so it's a little bit easier and not as challenging.

# NUMBERAMA WITH FAMILY AND FRIENDS: 2004 – 2006

## Overview

After my three year foundation grant funding for my Numberama program ended in 2004 (see below for a description of teacher responses to my Numberama program during this time), I secured funding from a few different local Maine school districts to continue engaging elementary school students with my Numberama problems and games. During this time period, from 2004 to 2006, I also worked privately with an 8-year-old mathematically gifted student, gave a perfect number lesson to my Creativity in Dementia author friends, and conducted a Family Math evening as a preview to my Numberama work at one of the schools that would be funding my program. In what follows, I am including a letter supporting my Numberama work with gifted children from the mother of the mathematically gifted child I worked with, who happens to also be the child that I gave the fictitious name of Edward to in my essay in Appendix 3. I am also including a letter supporting my Numberama work at a retirement home, from my Creativity in Dementia author friends. Finally, I am including a description of the very positive and appreciative responses that I received from my 2004 Family Math evening.

**Letter from Parent of Mathematically Gifted Child**: I am writing to express my unqualified support for Elliot Benjamin's Numberama program. Elliot taught my young son last year using his Numberama program. My son loved the experience. He looked forward each week to the day that he got to do math with Elliot and in between meetings delightedly worked independently on the projects and concepts that Elliot had introduced to him. At one point he asked Elliot if they could meet every day instead of once a week—a testament to how engaged my son was in Elliot's teaching.

I was impressed by Elliot's manner in working with a young child. He was patient and gentle yet disciplined and firm when necessary. Elliot Benjamin's creative Numberama program gives children a fun and engaging experience with numbers in which they learn fundamentals of mathematics along the way. I highly recommend his program and am happy to answer any questions. Please feel free to contact me if you wish.

## Letter in Support of Numberama with Seniors from Creativity in Dementia Authors/Friends

Elliot Benjamin has a gentle, caring and connective way of communicating with seniors through his Numberama project. This project invites people to challenge and explore numbers in a safe and fun atmosphere. This project offers an innovative and mind-sharpening program for seniors. Numberama offers a group activity that goes beyond the typical retirement and nursing home programming—enticing intellect and creating group energy.

## Family Math Night Numberama Responses

*Were these family night activities helpful to you? If so, would you tell us how?*

18—Yes 1—No

–mentally stimulating
–informative
–family togetherness
–playful approach to math
–understand what my son is doing in school
–tells us about new things that we didn't know
–reviewing patterns is always good
–great for the kids to realize that math can be fun
–it got my mind working and thinking again
–tricks to help me in school
–fun looking at patterns
–patterns are very helpful and thus time-saving
–easy to understand the concepts
–I found out math is even more confusing than I thought
–relaxing, but challenging
–learned an easier way to divide
–learned new things
–fun way of learning different kinds of math that I would not have at school

**What did you learn that will be helpful to your children with their schoolwork in the future? Did the activities increase your interest and your children's in math?**

–exploration of possibilities—push beyond right/wrong
–patterns to figure out perfect numbers
–encourage to try and keep trying
–simple ideas lead to patterns
–patterns are fun for everyone
–perfect numbers, edges, points,
–focus and have fun
–gave a new way to look at numbers and patterns
–algebra terms
–he'll have to do his own homework
–different games to help teach skills
–refreshed my memory in math
–look for patterns
–tricks and fun stuff
–I have always been interested in math and hope my child's interest will increase more
–look for patterns and concepts in the math problem

## NUMBERAMA AT AN ELEMENTARY SCHOOL IN REGULAR CLASSES: 2001 – 2004

### Overview

For the four school years from 2001 through 2004, I engaged my Numberama program with elementary school students in grades 3 through 6 at Mt. View Elementary School in Maine;

from 2002 through 2004 my program received grant funding and I visited various classrooms on a regular basis once a week, after voluntarily visiting a few classrooms once a month in 2001. The following teacher responses describe the stimulating and beneficial effects that my Numberama program had on the children in these classes.

**Teacher #1**: Dr. Benjamin's Numberama Program was a great supplement to our district's math program. Throughout the school year, I often noticed students transfer what they learned about number patterns with Dr. Benjamin to their everyday work in math. Dr. Benjamin was successful at engaging the majority of my students in experiences requiring hypothesizing and testing predictions.

**Teacher #2**: Dr. Benjamin's Numberama program challenged my students to use what they knew in order to work beyond traditional mathematics. They were always able to quickly pick up from where they left off on his previous visit. Students enjoyed finding perfect numbers.

**Teacher #3**: "Numberama," presented by Dr. Benjamin was of great value Grade 3/4 students. The children learned about patterns, mathematical terminology, and "thinking like" a mathematician. Most importantly, the students learned that "Doing Math" can be fun and can become a life-long recreational activity!

**Teacher #4**: The numberama program taught by Dr. Benjamin has been fun for the students in grade five. The students are challenged to think differently about math solutions. Their work has demonstrated that they can do more sophisticated math then they expected of themselves. The challenge and the motivation resulting from their own personal success has been good for them.

**Teacher #5**: This letter is written in support of the grant submitted by Dr. Elliot Benjamin. Dr. Benjamin has voluntarily worked with my sixth grade math students once a month. The activities he does with my students offers them an enrichment experience that I would not be able to provide my class any other way.

The students are interested, actively involved, and challenged by the activities he chooses to do with them. It would of great benefit to my incoming students to continue the work that Dr. Benjamin started with them in fifth grade.

**Teacher #6**: This letter is in appreciation of the efforts of Dr. Elliot Benjamin in visiting my 5th grade classroom to work on number theory, sets and other mathematical concepts. All these concepts fit in with what we are trying to cover in our classrooms anyway. In working on sets, my students grew familiar with concepts that have helped them understand division, fractions, percents, and probabilities.

His activities dealing with prime and perfect numbers have added greatly to their understanding of division, multiplication, greatest common factor and possibly other areas yet to come. The students really enjoy "playing" with numbers, and they also appreciate hearing a different voice and style of teaching. All in all, the program is very worthwhile, and I wish they could have the benefit of this expertise once a week instead of once a month. It is a well-spent hour in our program.

## NUMBERAMA IN MATHWORKS WORKSHOPS: 1990 – 1993

## Overview

The Mathworks (Numberama) workshops that I conducted in Belfast, Maine from 1990 to 1993 was designed primarily for elementary school and middle school teachers actively engaged in teaching mathematics. The responses of the teachers were generally quite appreciative and enthusiastic. In the two student responses that follow, the first student was an adult GED teacher who found my Mathworks/Numberama problems to be useful not only for the more advanced students, but also for the slower students who needed more practice in the basics.

The second student went way beyond anything I could have expected in a Mathworks participant's responsiveness, as she included a detailed description of how she applied various Numberama problems that I presented in her three levels of Gifted students that she worked with, including a number of informative attachments about Magic Squares history and constructions, and teaching techniques that she utilized. However, I am not including her Magic Squares attachments, since what is most relevant here is to see how a teacher working with gifted students responded to my Mathworks/Numberama program.

**Student #1**: I was very pleased to be allowed to take the Mathworks courses. Since I teach GED for adult ed. I come into contact with many adults with little knowledge of basic math and who are intimidated by math.

What I learned from the class is there is a way to make math fun but also give students a way to practice some basic skills while having fun. The concepts presented offer the slower students needed practice in basics but also offers the brighter students a challenge by figuring out whether there was a pattern involved in finding answers.

The two things that I could use in my classes are the information about the sums of squares and the Farey fractions.

**Student #2**: The Mathworks workshop is an ideal mathematics course for teachers of elementary math students, and especially for those who work with advanced math students. G/T teachers, working as they often do outside the classroom with small study-groups ranging in size from four to eight students, must provide interesting materials that will motivate gifted learners. The purpose of the small-group format is to enable the teacher to "push" the students, challenging them at the upper limits of their thinking abilities. The content for the specialized group is not the traditional grade-level math curriculum, but a potpourri of math activities that emphasize the conceptual level of mathematics. Following is a brief analysis of ways in which I've used Mathworks topics in my program for students who are mathematically gifted and/or intellectually gifted.

## G/T Math, Level 4

Topics most useful at this level were Clock Arithmetic, Fibonacci Numbers, Magic Squares, Pascal's Triangle, and Triangular Numbers.

Given enough preparation time, a teacher can design problems at progressively more challenging levels, of course, to that they can be used at different ability levels. I was fortunate in locating just such a progression of Magic Squares activities. . . . Beginning with the relatively simple Magic Circles, Level 4 students were able to complete Magic Triangles with ease. Magic Stars was much more difficult, even for some Level 5 students.

After completing Magic Squares, I challenged Level 5's with Pure Magic Squares: Create a magic square using the even, consecutive digits 10 – 26. The square in the center must be numbered 18. Note a: Most students thought Magic Square Formula. . . was "too easy." Note b: Many students, even gifted students in Level 6, become frustrated when asked to create their own magic squares. They can do it, but they don't want to/don't enjoy it.

## G/T Math, Level 5

Favorites at this level, in addition to Level 4 activities, are Abundant Numbers, Perfect Numbers, and Weird Numbers.

The most successful lesson at any level was the Fibonacci Numbers lesson with Level 5 students. The only explanation I can offer for the enthusiastic response to this lesson is that I introduced it as a "mystery of science,", as something that we know to be true in nature yet do not know how to explain. After giving several examples (pine cones, daisies, sunflowers, star fish, sand dollars, *etc.*), I used the "pairs of rabbits" problem. . . . The students were, in a word, fascinated. This week, I'll learn whether the excitement was enough to cause them to complete a homework assignment and return it a week later.

## G/T Math, Level 6

Most students that I work with have not reached Level 6, but I did try an extension of Fibonacci Numbers with advanced students in grade 5 who are intellectually gifted as well as mathematically gifted. Not entirely successful! Beginning with a warm-up activity using domino pieces. . . . and Domino Digs, I then introduced the table shown as Attachment 6. Of six students, only two were interested enough to complete the table as homework and bring it in the following week. Possible explanation: too much of a good thing; "boring" to them because they knew the patterns so well, knew what to expect. . . G/T students have a very low tolerance for repetitive work.

## Summary

In summary, this workshop provided a large number of ideas that I can use with my students. This will be evident, I think, after a quick review of Attachment 7, an outline of one of my Level 5 study-groups. Several activities done in the workshop were used directly to meet cognitive and affective objectives listed on this outline, especially cognitive objectives 1 and 2 and affective objectives 1, 2, and 4.

* The "Levels" refer not to grade-levels, but to levels of difficulty. Most study-groups are multi-grade.

## NUMBERAMA IN MY TEACHER EDUCATION COURSES AT UNIVERSITY OF MAINE: 1990 – 1993

### Overview

I experimented with various problems from Recreational Number Theory that were extracted from Chapter 1 of this *Numberama* book, in the course of my teaching the MAT 107-108 course sequence for future elementary school teachers at the University of Maine from 1990 to 1993. In my first semester as an instructor in this course, I quickly realized that the traditional exam structure for the course, consisting of numerous in-class unit exams, cumulative exams, and a final exam, was perceived as debilitating, oppressive, and laden with fear for practically every student in my class. I made a decision to change the exam structure drastically, instituting three take-home exams plus numerous "special problems" as described in the Numberama Recreational Number Theory problems in Chapter 1 of this book. The responses of the class were for the most part one of great relief and appreciation.

In this section of the Appendix, I am including 16 verbatim responses from students over the course of the six semesters that I have taught these classes. I believe that these accounts speak better than anything I can describe about how my students responded. However, I would like to emphasize a turning point in my approach to teaching the class that occurred during the second year of my teaching this course sequence. To my surprise and disappointment, a number of students found the "special problems" to be rather anxiety producing and frustrating, as they labored for hours upon end to find the solutions to these problems, while panicking that their grade would suffer for not coming up with the correct formulas. Although there was still considerable appreciation for the value of these problems, the conflict and concern about the negative responses motivated me to decide to minimize the required number of these problems in the second semester and offer most of them as bonus problems. For required problems I removed the emphasis on the grade penalties for not correctly solving a problem, stressing that what I was really looking for was an honest attempt to work on the problem.

Although I gave a fewer number of total problems over the semester, I found myself developing the problems more informatively and more extensively in class when I both introduced the problems and subsequently went over them. As can be seen from some of the verbatim accounts that follow, the student responses were quite dramatic in the transformation of their previous anxieties into a playful, appreciative, enthusiasm toward working on these problems over the second semester. Some of the students did research on their on their own in the library, reading more about Number Theory and the mathematicians who originated some of the theories.

A modest number of my students expressed a strong interest in purchasing my *Numberama* book in order to utilize my Numberama techniques of Recreational Number Theory with their own eventual students. However, as can be seen from these student accounts, although the great majority of the descriptions are very appreciative and positive, there were a few students who had much difficulty overcoming the anxiety that they felt toward math, and did not have a positive reaction to working on my Numberama problems.

I believe an important variable that made my presentations more effective the second year I taught the class was my consultant work in the Belfast Area School District, where I was working in a number of fifth and sixth grade classrooms, utilizing the very same Recreational Number Theory problems that I was presenting to my class at University of Maine. Over time the ideas, problems, and techniques gradually took shape in both of my settings, and made me confident that Numberama and Recreational Number Theory has definite concrete value with children—at least from my own experiential perspective.

## Student Responses

**Student #1**: I found your special math problems a challenge. They made me do a lot of deep thinking and figuring. The ability to solve problems is a must in today's world. It's much too easy to use a machine to get the correct answer. With your problems, I had to do all the work using my math skills. It was challenging and frustrating, but at the same time, it was rewarding when I got the correct answer. These have definitely given me a new outlook on math. A challenge is good for everyone and these can be used with kids who need to be challenged to show the joy of math.

**Student #2**: As a result of the special problems, I've learned that math is not the black and white science that I thought it was. I am a person who thinks in terms of abstracts and possibilities. My confidence in my math abilities has always been low. However, seeing that there are various ways to approach a problem proves that math isn't always an exact right or wrong situation. Solving the special problems has involved creativity and thought. There were no specific formulas. The old perceptions of memorizing formulas and doing busy work haven't held true. It has also been helpful to approach things from a child's perspective, reshaping our own way of thinking. There may be hope for me and math after all.

In addition, rather than taking strategies for granted from rote memory of childhood, we were able to learn why the strategies are true. The methods were explained, answers were proven. More than just the formula and answers—we learned the why. Finally, the background to what we always learned helped make the picture complete in logically understandable terms.

**Student #3**: I feel that Math 107 has broadened my views on the concept of math. Over the course of the semester, the math curriculum took on ideas such as problem-solving in chapter one and special problems which were assigned weekly. I found the problem-solving extremely difficult, especially at first. However, the class enthusiasm as well as that of the professor, helped to coax me into at least "getting my feet wet" through trying my best with a somewhat more positive attitude. The special problems seemed easy when seen in class, and they sparked a new way of thinking more openly to new mathematical ideas. I admit, some seemed fun and I am now able to visualize enjoyment that students, at varying levels would get from these. Even though I was very hesitant in trying out new ideas, I gained a more open-minded feel of new concepts. Whether or not I may one day pass on these ideas to my elementary class, I have gained further insight, and am perhaps better able to teach from having been exposed to them.

**Student #4**: When I found out I needed a science or math course to fulfill my degree requirements, I flipped a coin and it turned up heads which meant a math course was on my schedule. My expectation of a study in mathematics was a semester full of fear and dread because I hadn't studied math in a formal way for over thirty years. This fear subsided as prior knowledge was rekindled by the step by step presentation of new methods to problem solving. New knowledge was presented with clarity and systematically that it made problems not as difficult or frightening as I would have found them in past.

Especially helpful were the special problems which we as a class had to grapple with. They were very thought-provoking and at times frustrating until bells and lights of understanding went off leading to the probably solution of the problems. The probable solutions in my case were not always correct. They were, however, near enough to being correct so that when the correct solution was given in class the bells and lights of new understanding would really go off. Insightful learning is often times its own reward.

These problems were also every useful when I substitute taught a sixth grade math class a few times. I presented the special problems of perfect numbers and Pascal's Triangle and discovered they really appreciated the change of pace. The amount of attention and energy most of the students applied to these problems indicated to me that learning math doesn't always have to come in a traditional package.

This class has given me a new perspective of mathematics; that studying math doesn't have to be a fearful or dreadful experience, but is an insightful experience leading to new methods of problem solving.

**Student #5**: I think the special problems were a worthwhile addition to the material that we covered in this class. Each special problem required logical thinking and a systematic approach to find a pattern that fit the requirements. This helped me to organize my efforts toward a specific goal. They also gave me an insight into how mathematicians develop a theory that applies to all possibilities. It was a challenge to search for a key to these problems and then interpret what was discovered. This took more effort than simply reading a proof in the text but it was also more profitable. I found some of the proof in our text difficult to understand because of the complex way that they were presented. The special problems made the class more interesting but their main value was to create a method of thinking that was not prejudiced by preconceived ideas.

**Student #6**: My mind is again fuzzy after late nights working on math. Working on Geometry proofs however doesn't give me the satisfaction that long hours working on the special problems do. When I do a Geometry proof it brings back unpleasant memories of High School Geometry and never knowing (or caring) where to start with a proof. If I could look at proofs in the same way as special problems, as a challenge that is exciting and keeps pushing me to find an answer, it would be more fun than drudgery.

It's that challenging aspect that appeals to me. Also, as in the magic square problem, if you do figure out a solution it can spark further interest. I know I felt quite humbled after all the work I put into finding the solution, only to find through research that there were many more ways to solve the problem.

I think, as a future teacher, that the game-like challenge of special problems can be utilized to spur children to increase their learning ability. However—I hope my students won't stay up too many nights until 2 and 3 a.m.

**Student #7**: By using magic squares in class, I feel I can motivate children. First of all we

could do one on the board as a class. I would do the 3. Then I would encourage the children to look for a pattern (horizontal, vertical, and diagonal—all equal 15). I'd ask if they noticed anything else, for example, 3 x 3 = 9, there are 9 squares total.

Then I'd break the class up into cooperative groups to do a 4 magic square. As I want unity in the class, whichever group got the solution could choose a special activity, *i.e.*, reading by teacher for the last 15 minutes of class.

I would encourage the children to work on this during a couple of math classes and at home. If they come up with a magic square, I'd then ask them to record any patterns they find and share these with the class. Extra credit will be given to the class for observations.

I figure this project could be part of math classes during a week's time. Hopefully someone will do a 5 (or more) magic square. If not, I plan to show 5 and 7 and then have the students record and share their observance and then I will share what my college classmates came up with. I'd also do more research on my own beforehand, looking for other help.

Since I've started this class, I have been checking out the bookstores for math books. I'm always on the outlook for some tidbit that might entice a child to probe further, which is what you have certainly done with our class.

**Student #8**: I really wish I could say I enjoyed the special problems, but I didn't. In our first class you spoke a little bit about math anxiety. Needless to say I suffer from this syndrome. I had to struggle to complete my homework and the added stress of doing a problem that to say the least was difficult was a little more than I felt comfortable with. I can understand your rationale for having us do the special problems. To stretch ourselves, maybe even to put some fun into math, but when you are having a hard time just to understand the basics it's hard to enjoy the recreational side of mathematics.

I did enjoy the format of the class. The reviewing of each week's homework helped me to understand it better. I feel the take-home tests were a benefit to me. If a test is a learning experience, these tests were successful. When I would take the tests, I would spend hours upon hour reading and re-reading the textbook in order to get the correct answers. I feel I learned as much from the tests as I did from any other aspect of the class. If the tests had been given in class, it might have showed that I didn't know instead of helping me to learn what you felt was important for me to know.

I wish I could say that I was looking forward to Math 108, but I'm not. I can say that I know I will learn more about math and it will enable me to teach it better.

**Student #9**: I found the special problems are very challenging and interesting and sometimes enjoyable—if I am not too overwhelmed for time and don't get too frustrated with them. However, because they were for credit in this course, I found myself devoting a great deal of time using problem-solving techniques to find the answers to the special problems. I would often spend many hours guessing and testing and not come up with a solution to the problem. I would then spend many more hours looking for a pattern that wasn't there, or that I could just not see.

On one particular problem, I threw the book in the corner in frustration and went and soaked in a hot tub full of bubbles in which I searched my brain for a pattern and it came to me while I soaked. However, a lot of time had been spent on this one problem when I finally "saw the light!" Although I was pleased to have arrived at the answer, I was disturbed to think my relaxing bath was interrupted with math homework!

I found myself spending many hours on special problems that I should have spent on my homework and textbook because there were a lot of things I did not understand in class and should have spent more time working on it at home instead of doing the special problems. Maybe we should get credit for doing homework assignments and bonus points for doing special problems.

**Student #10**: To my understanding the special problems were designed to help develop thinking processes; to help one become aware of the relationships between numbers, formulas, theories; to build curiosity; and to have fun with puzzle-type problems. These appear to be valid objectives.

However, in my opinion, the problems were more than challenging (at least for me) and required hours and hours of figuring. Although I handed in every problem and for the most part, received full credit, I found these problems to be extremely difficult and often wondered how they were benefiting me directly. The grade points earned in no way reflect the amount of time spent on these problems. If such time-consuming and thought-provoking problems are going to be assigned, they should merit more worth. And yet, for those who do not have the hours, fortitude and thinking skills to manage these problems, they should not be penalized.

I suppose that on the average most people enjoy brain-teasers and puzzles and that they enjoy working on such things. I, on the other hand, have a very difficult time with puzzles and such and must work extremely hard to do the simplest kind of puzzles. They are not fun for me, but rather sheer drudgery. Personally, I would have enjoyed the course more without them. However, they represent the nature of math and for those who are inclined, I guess they do build curiosity and make math fun.

I would probably consider these special problems suit your objectives well.

**Student #11**: I felt the special problems made math interesting. It provided a break from the routine of whatever subject we were on. I think it's a wonderful idea to use with kids; it may spark someone's interest in math. The problems seemed like games or challenges. I also felt pressure was taken off by not making the special problems mandatory. Students could feel positive about them and therefore have more interest in doing them. Some parts of math need to start being fun so that more students will develop an interest. Children can get some positive reinforcement by doing the problems through the system of bonus points.

**Student #12**: Re: Math 108: I felt the special problems were much less stressful this semester as they were mostly used as bonus and not required for credit. Although I still attempted to solve the problems, I did not feel so bad when I did not find a solution even though I spent a lot of time on it. I did not feel so compelled to spend a great deal of time and frustration finding a solution especially when I drew a mental block as to where to even begin!

In all honest, I have to confess that if I hadn't had last semester to compare it with, I probably would not have spent as much time as I did on the special problems simply because they were not always required but were often considered only as bonuses. My human nature is to do my best of what is required of me only. I still do not consider math as recreational although I do not have the anxiety I had at the beginning of MAT 107.

**Student #13**: I found that the special problems were very challenging. Sometimes Math can be intriguing and gameish. These problems really helped me see mathematics in a different light. It was fun! I really like the idea that the value of the problems were bonus. The pressure of doing the special problems was alleviated and made it much more enjoyable to do. Number theory is a really neat part of math. I enjoyed seeing the patterns in many of the numbers (Fibonacci, *etc.*). This is something that students can really relate with and at a variety of levels. I am very interested in employing this into my future classroom in order to challenge students and give them more insight into mathematics.

**Student #14**: My reaction to the special problems has changed since last semester. Before, I had mixed feelings toward the special problems. I think they were interesting and important. I found myself going to the library more to find out who Fermat was or what a Diophantine number was. As a future teacher who will be teaching mathematics, I think it's important to know all aspects of mathematics that include the history of mathematics as well as the content of mathematics.

Although I may not have been successful at some of the special problems this semester, I did get some value out of them by doing a little reading about some of the mathematicians (*i.e.*, Euclid, Fermat, Cioanthus, *etc.*). Our UM library has a good selection of books on number theory. I was up there one night looking for Diophantine numbers, and I started browsing some of the number theory books. This was something I have never done before. This summer I hope to go up and do some free reading.

To conclude, I guess I did get a learning experience out of the special problems. I think they are useful. I think you should encourage reading about number theory in our next MAT 107-108 series. In addition to the special problems, I enjoyed the history of it.

**Student #15**: RE: MAT 108. It is the opinion of this student that the special problems were presented in a more relaxed manner than in MAT 107 last semester. In MAT 107 it seemed that we were assigned a special problem almost every week. Because of the frequency of the special problems, they sometimes overshadowed our homework assignments. I felt this added to the stress of an already stressful experience, for a student uncomfortable in math.

This semester the special problems were less frequent and in my opinion better presented. It seemed to me that you took more time setting up the groundwork for the special problems.

I've really enjoyed both 107 and 108 and hope that I can someday help students that I will have in the future to understand math as much as you have helped me.

**Student #16**: For me the special problems have been an excellent way to improve my problem solving abilities.

I have saved each one and hope to use them again in a class of my own some day. I like how they force a person to use basic skills over and over. Sometimes special problems bring unneeded or unwanted anxiety, but since the problems we had were not for core grade but extra credit it was more of a fun challenge.

Special problems also gave me a great feeling of accomplishment when I finally solved it. All in all it has been a great math class.

## NUMBERAMA IN MY FINITE MATH CLASSES AT UNITY COLLEGE: 1990 – 1993

### Overview

In my first two years as a mathematics professor at Unity College in Maine, from 1985 to 1987, my students were for the most part not responding very positively to my teaching, and my enthusiasm for teaching the standard Algebra and Pre-Calculus courses that I was assigned to teach was not particularly high. It was around this time that I became interested in Recreational Number Theory (see the Acknowledgments and Introduction sections in this book) and started experimenting with what I designated as Numberama problems with my 7-year-old son Jeremy (see the Introduction to the Games section). It was also around this time that I decided to try out these problems by developing and teaching Finite Math at Unity College, which came about from Stephanie Pall (see the Acknowledgments section) conveying to me that my tremendous enthusiasm when I talked about number theory was something that I should find a way of conveying to my Unity College math students, when I told her about my negative student evaluations).

Finite Math is a standard Liberal Arts major math course at colleges, but my inclusion of my Recreational Number Theory (Numberama) problems contributed quite the "non-standard" twist to the course. I continued teaching my Finite Math courses nearly every semester for the rest of my career as a mathematics professor at Unity College, until my retirement in 2006. The student responses to my Recreational Number Theory problems (which I referred as RNT problems) were for the most part very positive, and this enabled me to eventually get promoted to Associate Professor at the college. During the early years of my Finite Math teaching, from 1990 through 1993, I offered students some bonus points for submitting a candid description of their experiences with these RNT (Numberama) problems in Finite Math. The following is the responses I received from seven students in my classes.

**Student #1**: Elliot, I have honestly found this course to be enlightening. The material you chose and your patience with the class as "we" stumbled through number theories and infinite prime numbers is commendable. And even though I will probably never have the opportunity to use these systems and theories in the real world, I at least now know where they came from. I personally have found the material to be challenging. In fact I can say from personal experience that the people at Tylenol were enjoying the class also. The material was not hard, it just sometimes required some serious thought to bull my way through it.

The way the class is structured I feel was also beneficial, and allowed each student to think about math in a totally different light. Even the group sessions promoted teamwork and created a reservoir of ideas to solve a problem.

I think that there is even the potential for another Finite Math at the 2000 level. This course could include a more in-depth look at the creators (discoverers) of certain systems and theories, and a term paper on a topic that was not covered in the class.

Elliot, I would have written this regardless of the extra points. I do honestly feel that a 2000 level course may be in order. Thank you, Elliot, I did learn something.

**Student #2**: I found the mathematics class to be a rather enjoyable one. I am not very fond of math, but I enjoyed the subject material in this class. Through this class, I saw that mathematics is not just a field of numbers made for no real purpose. Instead, I noticed a way to enjoy mathematics and to witness the endless field of numbers and formulas. By this I mean that math seems to be an endless array of sets of numbers, formulas, *etc*. Math is no longer a boring subject to me when I reflect on this class. I have learned in a way to enjoy and respect the beauty of numbers.

**Student #3**: I felt that the class was unique. Being able to work with other students was helpful. Other students gave me more insight on how to accomplish homework assignments. It made me feel good when I understood something and someone else didn't understand it. I was able to help friends understand it better and *vice versa*. The class was interesting in the beginning but somehow you lost my interest toward the end of the class. I think the reason for this was because it started to get extremely complicated. But what was neat was my other friends helped as well as what I learned in the class.

**Student #4**: Finite Math: Finite Math has been beneficial to me for brain power. At points in my homework. I would become stuck and would have to push through to find the criteria. This style of learning tends to be more exciting because we don't simply go home and do the homework by repeating and practicing what we have already learned. Rather, we have a puzzle or brainteaser to solve.

The classes were almost always upbeat and interesting and the guest teacher was great.

One problem I have was with the glare on the blackboard. The largest concern in many classes was with the speed that you taught. I couldn't keep up at times, and found myself buried in the notes.

I like the dot game that we learned and I actually stumped a student the other day on it. Perfect numbers was good, and I'd like to learn more about probability.

I have a feeling that being exposed to so many math theories will help me in the future, and will give me an advantage when I at least recognize some of what we learned.

I give high marks for style, preparedness, office hours, and especially enthusiasm and interest.

**Student #5**: I found Finite Math to be enjoyable. It was a laid back atmosphere, where the student could simply try and have fun figuring out the various problems. One criticism that I would make would be to get rid of the book. The lectures that used the book were boring and not much fun. The days that the book wasn't used were much better. The lectures without the book were at least $3^2$ times better. I also felt that the class was more interested and responded better on those days.

**Student #6**: After the first session of this class, I kept returning in order to challenge my own mind as well as gather ideas which I could put to use in my classroom. I found it very stimulating, since the math taught in the sixth grade is not very challenging. I enjoy math and having to think.

I especially enjoyed the session with the games because I will be able to use those with my students at various times throughout the year. Thank you.

**Student #7**: When you first asked for introductions and reasons for taking the course the connection between "art" and math seemed vague. I come from a scientific family and everyone has a strong artistic background. Science and art are just objective and subjective views of the same world. Math keys in with this neatly and gives order and reason a relevance in art. I've always loved both the honest clarity of numbers and the geometry of nature. This course fitted my bill and whetted my appetite. Thank you.

# Appendix 2: Definitions, Examples, Hints

1. **Counting Numbers:** 1, 2, 3, 4, 5 …. continue adding 1 to each number to obtain the next number.
2. **Even Counting Numbers:** 2, 4, 6, 8, 10 …. continue adding 2 to each number to obtain the next number.
3. **Odd Counting Numbers:** 1, 3, 5, 7, 9 …. continue adding 1 to each number to obtain the next number.
4. **Multiples of** 7: 7, 14, 21, 28, 35 …. continue adding 7 to each number to obtain the next number.
5. **Triangular numbers:** 1, 3, 6, 10, 15, 21, 28 …. continue adding one more than the difference of the previous two numbers to obtain the next number.
6. **Fibonacci Numbers:** 1, 1, 2, 3, 5, 8, 13, 21 …. continue adding the sum of the previous two numbers to obtain the next number.
7. **Prime Numbers:** 2, 3, 5, 7, 11, 13, 17, 19, 23 …. prime numbers are numbers that have no divisors other than 1 and the number itself.
8. **Abundant Numbers:** numbers such that the sum of all the number's proper divisors, *i.e.*, all divisors other than the number itself, is greater than the number. An example of an abundant number is 24 since the proper divisors of 24 are 1, 2, 3, 4, 6, 12, and their sum is 28, which is greater than 24.
9. **Semi-Prime Numbers:** numbers that have exactly 4 divisors, including the number itself and 1. An example of a semi-prime number is 15 since the only divisors of 15 are 1, 15, 3, and 5.
10. **Kaprekar Numbers:** two-digit numbers such that their squares have four digits and the sum of the first two-digit number and last two-digit number of the square equals the original number. An example of a Kaprekar number is 45 since $45^2 = 45 \times 45 = 2025$ and $20 + 25 = 45$.
11. **Twin-Prime Numbers:** 2, 3, 5, 7, 11, 13, 17, 19, 29, 31, 41, 43, …. twin-prime numbers are prime numbers such that either the consecutive odd number above it is a prime or the consecutive odd number below it is a prime. We will consider 2 to be a twin-prime number since 3 is a prime. An example of a prime number that is not a twin-prime number is 23.
12. **Perfect Number:** a number such that the sum of all the proper divisors of the number equal the number itself. The first (smallest) perfect number is 6 since the proper divisors of 6 are 1, 2, and 3, and they add up to 6. The second perfect number has 2 digits; the third perfect number has 3 digits.
13. **Semi-Perfect Number:** a number such that some combination of the proper divisors of the number add up to the number itself. An example of a semi-perfect number is 36 since the proper divisors of 36 are 1, 2, 3, 4, 6, 9, 12, 18, and $18 + 12 + 6 = 36$.
14. **Weird Number:** a number that is abundant but not semi-perfect. Most abundant numbers are semi-perfect and therefore not "weird." However, there is one, and only one, two-digit weird number, and one, and only one, three-digit weird number.
15. **Amicable Numbers:** pairs of numbers such that all the proper divisors (excluding the number itself) of one number add up to the other number and *vice versa*. The smallest

pair of amicable numbers are both in the 200s.

16. **Prime Factorization:** means breaking up a number into a product of prime numbers with exponents. Examples of prime factorizations are $150 = 50$ x $3 = 25$ x $2$ x $3 = 5$ x $5$ x $2$ x $3 = 5^2$ x $2$ x $3$; $108 = 54$ x $2 = 27$ x $2$ x $2 = 9$ x $3$ x $2$ x $2 = 3$ x $3$ x $3$ x $2$ x $2 = 3^3$ x $2^2$. Prime factorizations are always unique: *i.e.*, a number can be broken up into prime numbers with exponents in one and only one way.

17. **Powerful Numbers:** numbers such that all primes in their prime factorization have exponents greater than or equal to 2. Some examples of powerful numbers are $8 = 2^3$, $9 = 3^3$, $100 = 2^2$ x $5^2$, $432 = 3^3$ x $2^4$. Note that 8 and 9 are consecutive powerful numbers.

18. **Set:** a collection of objects; in mathematics we will consider sets of numbers and denote it with parentheses. Some examples of sets are $\{1,2\}$, $\{1, 3, 5\}$, $\{2, 8, 9, 11\}$. Ø means the empty set—which has no elements.

19. **Subsets:** subsets are parts of sets. If the original set is $\{1, 2, 3, 4, 5\}$, some subsets are $\{1, 3\}$, $\{2\}$, $\{1, 2, 4, 5\}$. Note that the whole set $\{1, 2, 3, 4, 5\}$ is a subset of itself, and the empty set Ø is a subset of every set.

20. **Number of Subsets:** a set with no numbers, Ø, has 1 subset: Ø; a set with 1 number, $\{1\}$, has 2 subsets: Ø and $\{1\}$; a set with 2 numbers, $(1, 2\}$, has 4 subsets: $\{1\}$, $\{2\}$, $\{1, 2\}$, and Ø; a set with 3 numbers, $(1, 2, 3\}$, has 8 subsets: $\{1\}$, $\{2\}$, $\{3\}$, $\{1, 2\}$, $\{1, 3\}$, $\{2, 3\}$, $\{1, 2, 3\}$, and Ø. Continuing in this way a set with 4 numbers $\{1, 2, 3, 4\}$ will have 16 subsets. Note the pattern that seems to be unfolding for the number of subsets for a set of numbers.

21. **Sums of Squares:** (for prime numbers) some prime numbers can be written as the sum of two squares. For example, $37 = 36 + 1 = 6^2 + 1^2$ ; $29 = 25 + 4 = 5^2 + 2^2$. It is also the case that some prime numbers cannot be written as the sum of two squares: for example, look at 3, 7, 19, and 31. There is a way to determine very quickly whether or not an odd prime number can be written as the sum of two squares. Divide the odd prime number by 4; if the remainder is 1 then the number can be written as the sum of two squares, and if the remainder is 3 then the number cannot be written as the sum of two squares. Note that 2 is the only even prime number, and $2 = 1^2 + 1^2$.

22. **Number of Divisors:** means all the divisors of a number, including the number itself, and we denote this by the symbol "d". For example, d $(20) = 6$ since the divisors of 20 are 1, 2, 4, 5, 10, and 20 and there are therefore six divisors of 20. There is a relationship between exponents of the primes in the prime factorization of a number, and its number of divisors. Note that $20 = 10$ x $2 = 5$ x $2$ x $2 = 5^1$ x $2^2$. Some other examples are $d(7) = 2$ and $7 = 7^1$; $d(25) = 3$ and $25 = 5^2$.

23. **Syracuse Algorithm:** This says to do the following: take a number: if it is odd, multiply it by 3 and add 1 to it; if it is even, take half of it; continue the process. For example, if the number is 7, the sequence would be 7-2-11-34-17-52-26-13-40-20-10-5-16-8-4-2-1. Note that there are 17 numbers in this sequence, as after 1 the sequence would always repeat itself since we would have 1-4-2-1-4-2-1. The first billion numbers have been tested for the Syracuse Algorithm and it has been found that they always eventually yield 1.

24. **Multiplicative Persistence:** given a two-digit number, multiply the digits together; if you get another two digit number, multiply together again; continue the process until

you obtain a single-digit number. The number of steps it takes to get a single-digit number is the multiplicative persistence of the original number. Some examples are: 24-8, so 24 has multiplicative persistence of 1; 45-20-0, so 45 has multiplicative persistence of 2; 97-63-18-8, so 97 has multiplicative persistence of 3. There is one, and only one, two-digit number that has multiplicative persistence of 4.

25. **Anomalous Fractions:** These are proper two-digit fractions such that the second digit of the numerator fraction and first digit of the denominator fraction are the same, and when you cancel out these two identical digits you get an equivalent fraction to the original fraction, as can be seen by reducing both fractions to the lowest terms. For example 16/64 = 1/4 and 16/64 = 8/32 = 4/16 = 2/8 = 1/4 when reduced to the lowest terms.

26. **Pascal's Triangle:** Pascal's Triangle is the following triangle (Fig. **29**):

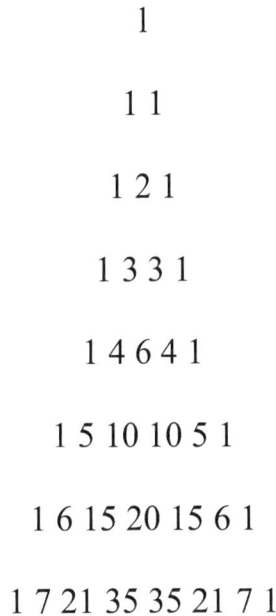

$$1$$

$$1\ 1$$

$$1\ 2\ 1$$

$$1\ 3\ 3\ 1$$

$$1\ 4\ 6\ 4\ 1$$

$$1\ 5\ 10\ 10\ 5\ 1$$

$$1\ 6\ 15\ 20\ 15\ 6\ 1$$

$$1\ 7\ 21\ 35\ 35\ 21\ 7\ 1$$

**Fig. (29).** Pascal's triangle in appendix 2: definitions, examples, hints.

It is found by putting 1's on the outside of the rows and adding two consecutive numbers in a row and writing this sum in the middle space of the row below. Note that the counting numbers are obtained as the second diagonal of Pascal's triangle, and the triangular numbers are obtained as the third diagonal of Pascal's triangle: 1. 3, 6, 10, 15, 21, .... One may also note that the sum of the rows of Pascal's triangle are doubling: 1, 1 + 1 = 2, 1 + 2 + 1 = 4, 1 + 3 + 3 + 1 = 8, 1 + 4 + 6 + 4 + 1 = 16, *etc.*

27. **Clock Arithmetic (Figs. 30, 31):** In an ordinary 12-hour clock, we can do arithmetic as follows: 9 + 5 = 14 − 12 = 2, 8 + 7 = 15 − 12 = 3, 5 x 4 = 20 − 12 = 8, 8 x 6 = 48 − (3 x 12) = 12, 5 x 5 = 25 − 2 x 12 = 1. Notice that for most numbers on the clock, you will never get 1. However, 5 x 5 = 1, and we therefore say that 5 has a multiplicative

inverse, namely 5, meaning a number that when multiplied by the original number, yields 1. On a 7-hour clock, every number, other than 7, has a multiplicative inverse, in the following way: 1 x 1 = 1, 2 x 4 = 1, 3 x 5 = 1, 4 x 2 = 1, 5 x 3 = 1, and 6 x 6 = 1.
We can talk about multiplicative inverses on any number clock; there is a simple way of determining whether or not every number on a clock has a multiplicative inverse (see Chapter 1).

**Fig. (30).** Clock arithmetic figure #1 in appendix 2: definitions, examples, hints.

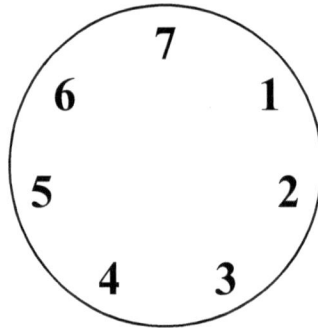

**Fig. (31).** Clock arithmetic figure #2 in appendix 2: definitions, examples, hints.

28.   **Farey Fractions:** these are sequences of fractions of a particular order; the sequence is going from zero to one, is consecutively increasing, all fractions in the sequences are reduced to lowest terms, and the fractions have denominators less than or equal to the order number. The Farey Fractions of orders 1 to 5 are as follows:
Order 1: 0/1, 1/1
Order 2: 0/1, 1/2, 1/1
Order 3: 0/1, 1/3, 1/2, 2/3, 1/1
Order 4: 0/1, 1/4, 1/3, 1/2, 2/3, 3/4, 1/1
Order 5: 0/1, 1/5, 1/4, 1/3, 2/3, 1/2, 3/5, 2/3, 3/4, 4/5, 1/1
Recall that to decide the order of magnitude of two fractions we obtain equivalent fractions using the least common denominator, as follows: 3/5 > 1/2 since 3/5 = 6/10 and

1/2 = 5/10 and 6/10 > 5/10. The distinctive characteristic of Farey fractions has to do with what happens when you cross-multiply two successive Farey fractions, *i.e.* multiply the denominator of the first fraction by the numerator of the second, and then multiply the numerator of the first fraction by the denominator of the second.

29.  **Magic Squares:** These are geometric squares divided up into boxes of the same size by constructing horizontal and vertical lines through the square, with a number in each box. The magical property of these squares is that the sum of the numbers in each of the rows, each of the columns, and each of the diagonals, all add up to the same number; this constant sum is called the "magic number." An example of a 3x3 magic square such that all the numbers can be seen from Fig. (**32**):

An example of a 4x4 magic square using all the digits from 1 to 16, such that each digit is used only once, with magic number 34 can be seen from (Fig. **33**):

An example of a 5x5 magic square, using all the digits from 1 to 25, such that each digit is used only once, with magic number 65 can be seen from (Fig. **34**):

There exists a 3x3 magic square, using all the digits from 1 to 9, such that each digit is used only once, with magic number 15. Can you find it?

| 479 | 71 | 257 |
|-----|-----|-----|
| 47 | 269 | 491 |
| 281 | 467 | 59 |

**Fig. (32).** Magic squares figure #1 in appendix 2: definitions, examples, hints.

| 16 | 3 | 2 | 13 |
|----|----|----|----|
| 5 | 10 | 11 | 8 |
| 9 | 6 | 7 | 12 |
| 4 | 15 | 14 | 1 |

**Fig. (33).** Magic squares figure #2 in appendix 2: definitions, examples, hints.

| 15 | 18 | 21 | 4  | 7  |
|----|----|----|----|----|
| 24 | 2  | 10 | 13 | 16 |
| 8  | 11 | 19 | 22 | 5  |
| 17 | 25 | 3  | 6  | 14 |
| 1  | 9  | 12 | 20 | 23 |

**Fig. (34).** Magic squares figure #3 in appendix 2: definitions, examples, hints.

# Appendix 3: Numberama and the Perfect Number Game with a Gifted Child

(August, 2008)

I will conclude this book with the following excerpt from an article that I wrote as part of a submission for my 2008 NECGT (New England Conference for Gifted Children) Numberama with Gifted Children workshop.

My article focuses on a very special gifted child, who I will refer to by the fictitious name of Edward, who initiated me into my 2007 - 2008 exceptional experience of teaching *Numberama* in the Old Town, Maine school district.

For many years I have conducted my Numberama program, the name I gave to various Recreational Number Theory problems and games included in my book: *Numberama: Recreational Number Theory in the School System* at a variety of school settings, family settings, community settings, and individual settings with children, adolescents, adults, and seniors. However, in the 2007-2008 school year I had the privilege and unique experience of offering my Numberama program as part of the Gifted & Talented services at the elementary and middle schools in the Old Town, Maine school district, working with between 45 and 50 children for 8 or 9 hours a week, comprised of approximately 15 - 20 children in the elementary school and 25 - 30 children in the middle school. Although I have successfully offered my Numberama program for gifted children in the past, both in schools and with individual students privately, the extensive and intensive Numberama experiences I have had with gifted children in every grade from grades 2 thru 8 this past school year have made me realize the enormous potential for the enhancement of mathematical creativity as well as for the collaborative and productive group participation in gifted children through the use of Recreational Number Theory. As a preview to my NECGT Numberama with Gifted Children workshop, I would like to give a brief illustration of my Numberama work with the gifted child, whom I will refer to by the fictitious name of Edward, who initiated me into my recent exceptional experience of teaching Numberama in the Old Town, Maine school district.

## A MATHEMATICALLY GIFTED CHILD

I first encountered Edward in March of 2005 when I was still a mathematics professor at a small college in rural Maine. One of my math professor colleagues asked me if I would be interested in working with the son of a friend of his, a mathematically gifted 5-year-old child, as my reputation for working with children through the mathematical exploration of number patterns described in my self-published book *Numberama: Recreational Number Theory In The School System* was well known in my rural Maine community. Although I was initially reluctant, as my *Numberama* work was geared towards children in grades 3 thru 8 who knew their basic arithmetic skills of multiplication and division, my colleague assured me that Edward was a highly unusual mathematically impressive 5 year old who was able to do multiplication and division of large numbers in his head. Thus my appetite was whetted; I agreed to meet Edward and try out a *Numberama* session with him.

I had worked with a number of mathematically bright children over the years, including children who were in the Gifted & Talented programs in various schools. But I had never encountered anything that even remotely resembled the mathematical abilities of little 5-year-old Edward who was only in Kindergarten at the time, and I am including children in grades 7 and 8 when I say this. Edward could indeed do all kinds of arithmetic in his head, including multiplication and division of numbers way into the thousands, and he even could do calculations with fractions in his head. His mathematical abilities were truly phenomenal, and he loved playing with numbers as much as I do. Edward was ripe for my *Numberama* lessons, and his mathematical creativity was truly astounding. Edward especially liked playing my *Perfect Number Game*, as his mother purchased my *Numberama* book and *Perfect Number Game* from me, and she and Edward would work on my *Numberama* problems and play *The Perfect Number Game* during the week. I would meet with Edward for an hour once a week, and give him suggested number patterns to think about and explore for the rest of the week. Edward became thoroughly immersed in the topic of Perfect Numbers, which is my favorite *Numberama* topic to teach (see my *Numberama* book for an illustration of how I teach this topic). Very basically, a perfect number is a number that is the sum of its proper divisors, *i.e.* you add up all the divisors of a number, not including the number itself, and if these add up to the number then the number is perfect. The first perfect number is 6, since $6 = 1 + 2 + 3$, and the second perfect number is 28, since $28 = 1 + 2 + 4 + 7 + 14$. Perfect numbers have a wonderful and surprising pattern to them, and there are some unsolved problems about perfect numbers that young children can readily appreciate, such as how many perfect numbers are there? (more than you can count?) and does there exist an odd perfect number? (see *Numberama*).

After discovering the pattern to formulate perfect numbers, we obtained the first five perfect numbers, which are 6, 28, 496, 8128, and 33,550,336 (see *Numberama*). Edward was captivated and intrigued, and I had a mathematical protégé whom I could not stop thinking about. This lasted for ten sessions, at the end of which Edward, who had just turned 6, had assimilated all he was able to of *Numberama* at that time. Summer was approaching, and Edward's mathematical interest was gradually lessening, as the more normal childlike parts of him that enjoyed riding his bicycle and playing outside with other children were taking precedence. We all agreed it was time to take a break from *Numberama*; we parted on very good terms with an openness to what the future might bring for us working together again.

Edward was a somewhat socially shy child, but he enjoyed playing with other children and he was not diagnosed with any kind of psychiatric mental illness that appears in the DSM-IV. The next time I encountered Edward was nearly 2 years later, as I was asked to work with a group of mathematically bright children in Edward's school. Edward was now in third grade (he skipped a grade) and was a few years younger than the rest of the children in the group. It was a group of over 20 very bright children who were mostly in grades 5 and 6, but Edward stood out as by far the child with the highest level of mathematical ability. However, it was apparent that Edward was rather quiet and not very comfortable socially with the other children, while he was bored in the class of children closer to his own age and had trouble with the rigid expectations of his teacher. Although I still would not classify Edward with any DMS-IV psychiatric syndrome, I was able to see that if these social difficulties were to continue, Edward could indeed be in danger of developing psychological problems as he got older. Edward took part in seven of our *Numberama* sessions with the older children, and I also worked with Edward individually for an additional five sessions, one of which was a joint session with a mathematically advanced child of the same age who was being homeschooled. It was apparent to me how much Edward enjoyed having the social and intellectual stimulation of another child whom he could relate to, but I also realized that this other child, although mathematically bright, was not on Edward's mathematical level, and it would not work mathematically to put them together in a group.

The purpose of my individual sessions with Edward was to teach him more traditional math, such as Algebra, in order to keep him stimulated mathematically. However, what Edward really wanted to do was more *Numberama*. In our last session I realized that the abstractness of Algebra was beyond the assimilation abilities of his cognitive 8-year-old mind in the Piaget context, and I yielded to his request to play a number game which involved using the different arithmetic operations (*i.e.*, addition, subtraction, multiplication, division) in creative combinations to come up with a positive number. We spent much of this last session struggling together to try to come up with one of these number puzzle problems that I do believe was an error. At one point when he was frustrated with a different number puzzle, Edward became very sad and tearful, and though I was able to help him work through this, I realized how delicate and vulnerable this 8-year-old artistically inclined mathematically gifted child truly was.

## SEQUEL

The sequel to my work with Edward is that he attended a wonderful little private school, Riley school lin Rockport, Maine, which happens to be the same school my own son attended for grades 7, 8, and 9. This school focuses upon nourishing a child's creativity within a social environment full of warmth, caring, and support, containing aspects of A.S. Neill's (1960) Summerhill philosophy and Carl Rogers' (1969) humanistic education philosophy. I think this is a perfect fit for Edward, and in addition to the supportive nurturing atmosphere of his new private school, Edward also has the good fortune of having the loving, caring, intelligent, and dedicated caretaking of his mother.

# BIBLIOGRAPHY

Adams, W.W., & Goldstein, L.G. (1976). *Introduction to number theory.* Englwood Cliffs, NJ: Prentice Hall.

Beiler, A.H. (1966). *Recreation in the theory of numbers.* New York: Dover Publications.

Brown, S.I. (1973). Mathematics in humanistic themes: Some considerations. *Educ. Theory, 3*, 141-214.

Brown, S.I. (1976). From the Golden Rectangle and Fibonacci to pedagogy and problem posing. *Math. Teach, 64*(3), 180-186.

Brown, S.I. (1983). *The art of problem posing.* Philadelphia: The Franklin Distribution Press.

Dence, T.P. (1983). *Solving math problems in basic.* Blue Ridge Summit, PA: Tab Books.

Dewey, J. (1933). *How we think.* Boston: D.C. Health and Co..

King, J.P. (1993). *The art of mathematics.* New York: Ballantine Books.

Miller, C.D., & Heeren, V.E. (1961). *Mathematical ideas.* Glenview, Illinois: Scott Foresman & Co.

National Council of Teachers of Mathematics (NCTM). (1981). *Mathematics assessment: Myth, models, good questions, and practical suggestions.* Reston, VA: NCTM.

National Council of Teachers of Mathematics (NCTM). (1984). *Curriculum and Evaluation Standards for School Mathematics.* Reston, VA: NCTM.

National Council of Teachers of Mathematics. (1991). *Professional Standards for teaching mathematics.* Reston, VA. NCTM.

Neill, A.S. (1960). *Summerhill: A Radical Approach to Child Rearing.* New York: Hart Publishing Co..

Postman, N., & Weingartner, C. (1969). *Teaching as a subversive activity.* New York: Delacarte Press.

Rogers, C.R. (1969). *Freedom to Learn.* Columbus, OH: Charles E. Merrill Publishing Co..

Walter, M.J., & Brown, S.I. (1971). Missing ingredient in teacher training: One remedy. *Am. Math. Mon, 78*, 394-404.
[http://dx.doi.org/10.2307/2316913]

Walter, M.J., & Brown, S.I. (1977). Problem posing and problem solving: An illustration of their interdependence. *Math. Teach, 70*(1), 4-13.

Wells, D. (1986). *The Penguin dictionary of curious and interesting numbers.* New York: Penguin Books.

# SUBJECT INDEX

www.ingramcontent.com/pod-product-compliance
Lightning Source LLC
Chambersburg PA
CBHW041728210326
41598CB00008B/815